SpringerBriefs in Applied Sciences and Technology

SpringerBriefs present concise summaries of cutting-edge research and practical applications across a wide spectrum of fields. Featuring compact volumes of 50 to 125 pages, the series covers a range of content from professional to academic.

Typical publications can be:

- A timely report of state-of-the art methods
- An introduction to or a manual for the application of mathematical or computer techniques
- A bridge between new research results, as published in journal articles
- A snapshot of a hot or emerging topic
- An in-depth case study
- A presentation of core concepts that students must understand in order to make independent contributions

SpringerBriefs are characterized by fast, global electronic dissemination, standard publishing contracts, standardized manuscript preparation and formatting guidelines, and expedited production schedules.

On the one hand, **SpringerBriefs in Applied Sciences and Technology** are devoted to the publication of fundamentals and applications within the different classical engineering disciplines as well as in interdisciplinary fields that recently emerged between these areas. On the other hand, as the boundary separating fundamental research and applied technology is more and more dissolving, this series is particularly open to trans-disciplinary topics between fundamental science and engineering.

Indexed by EI-Compendex, SCOPUS and Springerlink.

More information about this series at http://www.springer.com/series/8884

Ajay Giri Prakash Kottapalli • Kai Tao
Debarun Sengupta • Michael S. Triantafyllou

Self-Powered and Soft Polymer MEMS/NEMS Devices

 Springer

Ajay Giri Prakash Kottapalli
Department of Advanced Production
Engineering, Engineering and Technology
Institute Groningen (ENTEG), Faculty of
Science and Engineering
University of Groningen
Nijenborgh, Groningen, The Netherlands

MIT Sea Grant College Programme
Massachusetts Institute of Technology
Cambridge, MA, USA

Debarun Sengupta
Department of Advanced Production
Engineering
University of Groningen
Groningen, The Netherlands

Kai Tao
Northwestern Polytechnical University
Xi'an, Shanxi, China

Michael S. Triantafyllou
MIT Sea Grant College Programme
Massachusetts Institute of Technology
Cambridge, MA, USA

ISSN 2191-530X   ISSN 2191-5318   (electronic)
SpringerBriefs in Applied Sciences and Technology
ISBN 978-3-030-05553-0   ISBN 978-3-030-05554-7   (eBook)
https://doi.org/10.1007/978-3-030-05554-7

Library of Congress Control Number: 2018965413

This Springer imprint is published by the registered company Springer Nature Switzerland AG
The registered company address is: Gewerbestrasse 11, 6330 Cham, Switzerland

Preface

Recent advances in Internet of Things (IoT) and sensor networks reveal new insight into the understanding of traditional power sources with the characteristics of mobility, sustainability, and availability. Conventional technologies which employ batteries to supply power may not be the preferred choice. In this age of advanced smartphones and wearable devices, the need of unlimited power has become a basic necessity. Most of the gadgets rely on some sort of power source in the form of batteries or power adapters. For example, smart watches and phones have become very common these days and have a huge potential for implementation of energy harvesters. In near future, it will be highly desirable to have self-powered smart wearable devices which meet their energy needs by scavenging mechanical energy produced by physical activities. In order to solve the problem of fast battery depletion in modern smart devices, significant research has been carried out in the field of energy harvesters especially using MEMS/NEMS technologies.

Polymeric materials have gained tremendous recognition for applications in sensor technology replacing classical solid-state devices. This is due to the many advantages that polymers offer, namely, low-cost devices, simple fabrication techniques, ease of fabrication into different forms (films, sheet, nanofibers, nanowires, and nanoparticles). In addition, the physical and chemical properties of polymers could be tailored for a specific sensing application. Fabrication of soft material sensors using natural and synthetic polymers has grown rapidly in recent years. This is attributed to the increasing demand of flexible and wearable electronics or devices required in many fields ranging from biomedical devices, electronic skin, optoelectronics, and robotics applications.

This book presents a brief review of the MEMS/NEMS energy harvesters and self-powered sensors. This includes a critical analysis on various energy harvesting principles (electromagnetic, piezoelectric, electrostatic, triboelectric, magnetostrictive, etc.) and novel materials (PVDF, PZT, graphene, etc.) for MEMS energy harvesters. A special focus has been dedicated to the recent efforts in developing flexible and stretchable self-powered soft polymer sensors, which use hydrogels, electrospun nanofibers, graphene, and conductive and electroactive polymers.

Chapter 1 reviews the recent progress in kinetic MEMS/NEMS-enabled energy harvesters as self-powered sensors. Recent advances and challenges in MEMS/NEMS-enabled self-sustained sensor working mechanisms including electromagnetic, piezoelectric, electrostatic, triboelectric, and magnetostrictive are reviewed and discussed. Recent advances in Internet of Things (IoT) and sensor networks reveal new insight into the understanding of traditional power sources with the new characteristics of mobility, sustainability, and availability. Individually, the power consumption of each sensor unit is low; however, the number of units deployed is huge. As predicted by Cisco, trillions of sensors will be distributed on the earth by 2020. Conventional technologies which employ batteries to supply power may not be the ideal choice. Energy harvesting systems as self-sustained power sources are capable of capturing and transforming unused ambient energy into electrical energy. During the last two decades, intensive efforts have been made towards the development of micro-/nanoelectromechanical systems (MEMS/NEMS)-enabled energy harvesting technologies which yielded breakthroughs in self-powered sensor evolutions.

During the recent years, nanogenerators fabricated using polymers like PVDF which exhibit piezoelectric behaviour have been investigated extensively due to their low production cost and high conversion efficiency. Polymer-based nanofiber energy harvesters are not only relevant for wearable devices but also for biomedical energy scavenging applications primarily due to their flexibility and biocompatibility. Chapter 2 presents a comprehensive review of the ongoing research work in the field of flexible micro/nano energy harvesters and the various applications of such devices.

Chapter 3 presents a comprehensive review of the various biomimetic self-powered and low-powered MEMS pressure and flow sensors that take inspiration from the biological flow sensors found in the marine world. The sensing performance of the biological flow sensors in marine animals has inspired engineers and scientists to develop efficient state-of-the-art sensors for a variety of real-life applications. In an attempt to achieve high-performance artificial flow sensors, researchers have mimicked the morphology, sensing principle, materials, and functionality of the biological sensors. Inspiration was derived from the survival hydrodynamics featured by various marine animals to develop sensors for sensing tasks in underwater vehicles. The mechanoreceptors of crocodiles have inspired the development of slowly and rapidly adapting MEMS sensory domes for passive underwater sensing. Likewise, the lateral line sensing system in fishes which is capable of generating a three-dimensional map of the surroundings was mimicked to achieve artificial hydrodynamic vision on underwater vehicles. Harbor seals are known to achieve high sensitivity in sensing flows within the wake street of a swimming fish due to the undulatory geometry of the whiskers. Whisker-inspired structures were embedded into MEMS sensing membranes to understand their vortex shedding behavior. At the outset, this work comprehensively reviews the sensing mechanisms observed in fishes, crocodiles, and harbor seals. In addition, this chapter presents an in-depth commentary on the recent developments in this area where different researchers have taken inspiration from these aforementioned underwater creatures and developed some of the most efficient artificial sensing systems.

Groningen, The Netherlands   Ajay Giri Prakash Kottapalli
Xi'an, China   Kai Tao
Groningen, The Netherlands   Debarun Sengupta
Cambridge, MA, USA   Michael S. Triantafyllou

Contents

Chapter 1
MEMS/NEMS-Enabled Energy Harvesters as Self-Powered Sensors

Kai Tao, Honglong Chang, Jin Wu, Lihua Tang, and Jianmin Miao

1.1 Introduction

Recent advances in Internet of things (IoT) and sensor networks reveal new insight into the understanding of traditional power sources with the new characteristics of mobility, sustainability, and availability. This aspect has become more challenging in providing power to microelectronic devices as they have become smaller, wireless, portable, and varied in their range of applications. Research into the development of a sustainable power source has been a key focus in the past decade.

One notable application concern in IoT is the supply power for wireless sensor network (WSN) nodes. These WSN nodes are expected to be inconspicuous, small, and self-contained sensor nodes. They are deployed in abundance for monitoring

K. Tao (✉) · H. Chang
The Ministry of Education Key Laboratory of Micro and Nano Systems for Aerospace, Northwestern Polytechnical University, Xi'an, China

Unmanned System Research Institute, Northwestern Polytechnical University, Xi'an, China
e-mail: taokai@nwpu.edu.cn

J. Wu (✉)
State Key Laboratory of Optoelectronic Materials and Technologies and the Guangdong Province Key Laboratory of Display Material and Technology, School of Electronics and Information Technology, Sun Yat-sen University, Guangzhou, China
e-mail: jwu6@ntu.edu.sg

L. Tang (✉)
Department of Mechanical Engineering, University of Auckland, Auckland, New Zealand
e-mail: l.tang@auckland.ac.nz

J. Miao
Department of Mechanical and Aerospace Engineering, Nanyang Technological University, Singapore, Singapore

and data collection and transmission about the physical environment or conditions. WSNs have been deployed to monitor domestic environments, building structures and bridges, as well as military tracking and fire detections. Individually, the power consumption of each sensor unit is low; however, the number of units deployed is huge. As predicted by CISCO, trillions of sensors will be distributed on the earth by 2020. Currently, its proliferation in use is however constrained by the amount of power it can generate.

A typical block diagram assembly of the WSN nodes is shown in Fig. 1.1. Each node is mainly composed of four basic elements: a sensing component to detect the physical environment or conditions by microelectronic sensors, a data processing unit by ADC module, a data transmitting module and the power supply by batteries. The miniature sensor node can not only serve for environmental sensing and monitoring but also function as intelligent data mining and storing.

As of now, using chemical batteries is the main method to supply power for WSN nodes. However, as the amount of distributed sensor network grows and electronic devices become smaller in size, providing sustainable power by batteries represents a great challenge. This challenge is further compounded if WSN nodes are deployed in hazardous and nonaccessible areas where replacement of batteries would not be viable. Secondary power source in addition to or replacement of batteries would therefore need to be established. Meanwhile, the advancements in CMOS technology and micro-/nanoelectromechanical systems (MEMS/NEMS) technologies allow the WSN nodes and microelectronics to become even more miniaturized, low powered, and easy to integrate. Therefore, harvesting ambient energy to supplement or replace batteries as power source to operate low-power wireless electronic devices would be an invaluable alternative.

Figure 1.2 shows the new type of wireless sensor network nodes powered by green and sustainable energy harvesting from the ambient environment. Compared to the traditional WSNs as shown in Fig. 1.1, the new microsystem has three additional parts, which are the ambient energy source, energy harvester, and the buffer. Ambient energy sources are usually in the form of light, sound and radiofrequency (RF) waves, heat, and mechanical motions. Energy harvesting systems are capable of scavenging and transforming various ambient energies to electricity by photovoltaic, thermoelectric, piezoelectric, and electrostatic energy conversion approaches, etc. The electrical energy can be further stored in a buffer, such as supercapacitors

Fig. 1.1 Typical block diagram of the WSN nodes with battery as power source

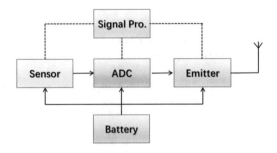

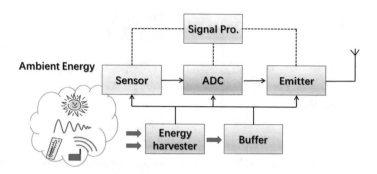

Fig. 1.2 Block diagram of the WSN nodes powered by green and sustainable energy harvesting from the ambient environment

Table 1.1 Table of wireless sensor states and associated power consumption [2]

Function	Duration(s)	Power(mW)	Description
Sleep	10^{-1} to 10^4	10^{-1} to 10^{-3}	A minimal power state allowing the device to "wake up" on event interruptions and power an internal clock
Polling	10^{-5} to 1	10^{-4} to 100	The power required to interact with the node's environment from sensing to actuating and whatever data processing are required on node
Transmitting	10^{-6} to 1	1 to 100	The power and time required to another node or a base station. This packet includes sensor ID, routing information, and sensor data value. The power required scales with the distance the data need to travel and the rate at which the information is sent
Receiving	10^{-3} to 2	10^{-1} to 1	The power and time required to listen for a data packet note that listening takes more time than sending to ensure data are received

or rechargeable batteries. The buffer is then used to perform a measurement cycle when enough energy is stored in the buffer. The energy circulation and data flow of the self-powered microsystem is schematically depicted in Fig. 1.2.

Table 1.1 summarizes the four main functions of a traditional WSN module, namely, transmitting, receiving, polling, and sleeping. To operate these functions in the microsystem, power consumption is in the range of microwatt level to milliwatt level depending on the distance and rate of data transmitting. If we consider a low-power temperature WSN module as an example, the power consumption of STLM 20 from ST Micro is about 12 μW at quiescent status. The ADC module from Sauerbrey requires 1 μW to perform an 8-bit sampling operation at 4 kS/s, whereas an IMEC transmitter consumes only 0.65 nJ per 16 chip burst operating at a low duty cycle [1]. As a whole, the power consumption can be lowered down to the level of tens to hundreds of microwatts, providing huge potentials for self-powered microsystems by continuously converting the ambient energy to the electricity.

The recent rapid development of IoT further requires the WSN to be more intelligent, multifunctional, and self-sustainable. The development of self-powered

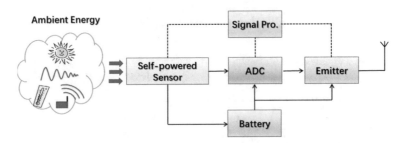

Fig. 1.3 New advancement of WSN nodes by using self-powered sensors serving as both sensing element and power source

active sensors by capturing and transforming unused ambient energy into the electrical energy would be the optimum solution. It does not need complicated energy management and coupling between the power source and sensing element. The advancement of micro-/nanoelectromechanical systems (MEMS/NEMS)-enabled energy harvesting technologies in the last decade has yielded breakthroughs in self-powered sensor research. MEMS/NEMS-enabled energy harvesting systems can not only perform as power sources to supply electricity to WSN nodes but can also simultaneously function as self-sustained sensors. A new advancement of WSN nodes by using self-powered sensors serving as both sensing element and power source is shown in Fig. 1.3. This further opens up new paths for future self-sustainable and smart sensing systems. This chapter mainly reviews the recent progress in MEMS/NEMS-enabled kinetic energy harvesters which function as self-powered sensors. Recent advances and challenges in MEMS/NEMS-enabled self-sustained sensor with energy conversion mechanisms based on electromagnetic, piezoelectric, electrostatic, triboelectric, and magnetostrictive principles are reviewed. The suggestions for the next generation of intelligent wireless sensing are also discussed.

1.2   Methods of Mechanical Energy Harvesting

There are several existing energy source candidates in the environment, which can be used for harvesting. The power densities of different types of energy sources are summarized in Table 1.2. Of these, solar energy is one of the most promising sources. Energy harvesting from direct outdoor sunlight is clearly the most effective one. Unfortunately, it is not the case for indoor lighting which operates at a level comparable with thermal and airflow energy. The performance of thermoelectric energy harvesting methods is constrained by minor temperature gradient over a short length, where energy can only be generated through temporal temperature difference using pyroelectric materials. RF and acoustic energy levels are several orders of magnitude less than other forms of energy sources. Among these,

Table 1.2 Comparison of energy harvesting methods in the environment

Energy types	Forms	Energy level	Conversion mechanisms	Reference
Solar energy	Sunlight	Outdoors: 150 mW/cm^3 Indoors: 10 μW/cm^3	Photovoltaic	[3, 4]
Thermal energy	Temperature gradient	5–100 mW/cm^3	Seebeck effect Thermoelectric	[3–5]
Acoustic noise	Wave	0.003 μW/cm^3 at 75 dB 0.96 μW/cm^3 at 100 dB	Piezoelectric	[6]
Radiation	RF signal	<1 μW/cm^2	Electromagnetic	[4, 7]
Fluid flow	Wind, ventilation, tidal wave, flowing water	Air: 200–800 μW/cm^3 Water: 500 mW/cm^3	Electromagnetic Piezoelectric	[3]
Mechanical motion	Structure and machine vibrations, human motion	4–800 μW/cm^3	Electromagnetic Piezoelectric Electrostatic	[8–11]

mechanical vibration energy is of good interest because of its versatility and ubiquitous characteristics. Ambient motions can be derived from structures, human body, vehicles, machinery, or air/water flows. Due to their pervasive existence, it is more suitable for small-scale embedded and deployable applications, which fit the goal of sustainable development in reducing the use of battery and performing intelligent monitoring.

Mechanical energy can be transformed to electricity by exploiting the mechanical strain or relative movement in the transducer. There are mainly four transduction mechanisms, namely, electromagnetic (magnetostrictive), piezoelectric, electrostatic, and triboelectric (Fig. 1.4). Electromagnetic transduction mechanism is based on the Faraday's law of induction where the electricity can be generated when there is a change of magnetic flux in a closed coil. The change of magnetic flux can be induced by two approaches: the relative movement of the magnet and the coil or *Villari* effect of certain ferromagnetic materials. Piezoelectric transduction mechanism mainly uses certain piezoelectric materials, such as lead zirconate titanate (PZT) and polyvinylidene fluoride (PVDF), where electric charges can be generated when the piezoelectric material is subjected to mechanical strain. Electrostatic transduction mechanism operates based on a variable capacitor under constrained bias voltage or charge. Depending on the external bias, it can be categorized into two types: electret-free electrostatic harvesters, where the constant bias is provided by external circuit, and electret micropower generators, where a precharged or dipole electret material provides the constant bias charge. There is another type of electrostatic energy harvester called as electroactive polymer-based harvester where the capacitance variation is induced through the deformation of certain electroactive

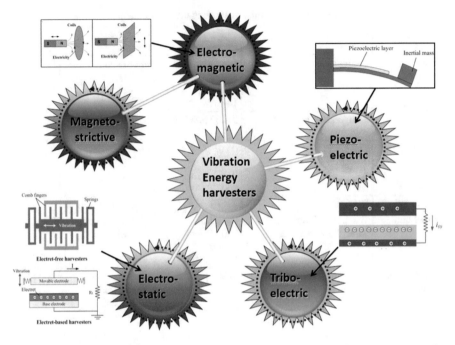

Fig. 1.4 Four main types of vibration-to-electricity transduction mechanisms: electromagnetic (magnetostrictive), electrostatic, piezoelectric, and triboelectric

polymers (EAPs). Most recently, Wang et al. [12] reported a new method of energy harvesting, namely, triboelectric nanogenerators, where charges can be created when two dissimilar materials come into contact with each other. Details of the principles and recent advances of these three transducers will be presented later in this section.

1.2.1 Electrostatic MEMS Harvesters

Electret-Free Electrostatic Harvesters

The basic principle of electrostatic vibration energy harvesting is based on the capacitance change of a variable capacitor under a constant bias. There are mainly two types of electrostatic energy conversion schemes depending on the different bias conditions, namely, voltage-constrained energy conversion cycle and charge-constrained energy conversion cycle. In the voltage-constrained cycle where the voltage is kept constant, the charge will circulate when the capacitance changes due to the relative motion between the two capacitor plates. In the charge-constrained conversion cycle where the charge is kept constant, the voltage varies with the capacitance due to the relative motion of the two capacitor plates.

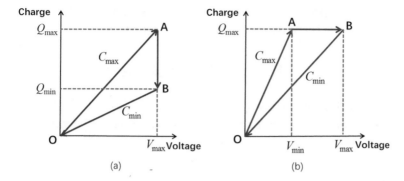

Fig. 1.5 (**a**) Voltage-constrained energy conversion cycle; (**b**) charge-constrained energy conversion cycle

Figure 1.5a shows the voltage-constrained energy conversion cycle in the QV chart. From point O to point A, the voltage in the capacitor is charged to the maximum value of V_{max}. At this moment, the energy stored in the capacitor can be calculated as

$$E_A = \frac{1}{2}V_{max}Q_{max} = \frac{1}{2}V_{max}\left(V_{max}C_{max}\right) = \frac{1}{2}C_{max}V^2_{max} \tag{1.1}$$

From point A to point B, the voltage in the capacitor is kept constant at V_{max}, while the capacitance changes from C_{max} to C_{min}. The charges stored in the capacitor discharge from Q_{max} to Q_{min}. To keep a constant voltage V_{max} at the whole system, the amount of energy stored in the capacitor would decrease from $Q_{max}V_{max}$ to $Q_{min}V_{max}$. This amount of energy originates from external mechanical source. From point A to point B, the capacitor is discharged to external circuit. The energy stored in the capacitor can be expressed as

$$E_B = \frac{1}{2}V_{max}Q_{min} = \frac{1}{2}V_{max}\left(V_{max}C_{min}\right) = \frac{1}{2}C_{min}V^2_{max} \tag{1.2}$$

Therefore, at the end of the cycle, the total converted energy from the mechanical source to electricity can be calculated as

$$
\begin{aligned}
E_{net} &= E_B + V_{max}\left(Q_{max} - Q_{min}\right) - E_A \\
&= \frac{1}{2}C_{min}V^2_{max} + V_{max}\left(C_{max}V_{max} - C_{min}V_{max}\right) - \frac{1}{2}C_{max}V^2_{max} \\
&= \frac{1}{2}\Delta C V^2_{max}
\end{aligned}
\tag{1.3}
$$

where ΔC is the capacitance change from C_{max} to C_{min}. The net energy converted from mechanical power source to electrical energy is equal to the blue-shaded area as shown in Fig. 1.5a.

Similarly, in the charge-constrained energy conversion cycle as shown in Fig. 1.5b, the total converted energy can be expressed as

$$E_{net} = \frac{1}{2} C_{min} V^2_{max} - \frac{1}{2} C_{max} V^2_{min}$$ (1.4)

The charge is kept constant at Q_{max} in the charge-constrained cycle as shown in Fig. 1.5b; the charge at point A and point B is identical and can be expressed as

$$Q_{max} = C_{min} V_{max} = C_{max} V_{min}$$ (1.5)

Therefore, the total converted energy can be rewritten as

$$E_{net} = \frac{1}{2} V^2_{min} \left(C_{max} - C_{min} \right) \frac{C_{max}}{C_{min}} = \frac{1}{2} V^2_{min} \frac{C_{max}}{C_{min}} \Delta C$$ (1.6)

The net energy converted from mechanical power source to electrical energy is equal to the red-shaded area as shown in Fig. 1.5b. It can be seen that the initial voltage needs to be charged to V_{max} at voltage-constrained energy conversion cycle, while the voltage only needs to be charged to V_{min} at charge-constrained energy conversion cycle. Due to the low voltage levels supplied by external charge conservation, the voltage-constrained type is limited to the low-power consumption applications. Roundy et al. [13] from Berkeley further categorized the electrostatic converters into three conversion types such as in-plane overlap varying, in-plane gap closing, and out-of-plane gap varying, as shown in Fig. 1.6. Although variable gap-type devices were less sensitive to the parasitic capacitance, these devices exhibited higher loss due to air damping and large surface area which might result in electrostatic stability problems. Therefore, an in-plane gap closing-type harvester was fabricated by deep reactive-ion etching (DRIE) process on the top layer of silicon-on-insulator (SOI) wafers. It was reported that an optimal output power of 116 µW/cm³ at a minimum gap of 0.25 µm could be obtained.

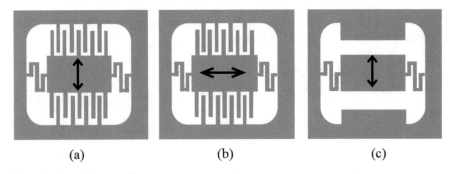

(a)   (b)   (c)

Fig. 1.6 Three different electrostatic conversion modes: (**a**) in-plane overlap varying, (**b**) in-plane gap closing, (**c**) out-of-plane gap varying

Electret-Based Electrostatic Energy Harvesters

One of the main challenges in electrostatic energy harvester is the requirement of external bias. Electret-based energy harvester employs an electret layer as its positive/negative permanent surface voltage bias. Electrets are insulating dielectrics with quasi-permanent electric charge or dipole polarization, which can be used to provide a biasing electric field, such as MEMS electret microphones. Generally, the electret-based energy harvesters usually operate in two ways: in-plane gap varying and out-of-plane gap closing. In-plane gap varying type transforms mechanical energy to electrical energy by changing the overlapping area. To enhance the performance with small displacement, the electret thin film and the electrodes are usually patterned with interdigital or other micro-sized shapes. The operating principle is shown in Fig. 1.7. When the top plate is excited and gives in-plane oscillations, the capacitance varies and generates charge circulation in the external circuit. With the interdigital-patterned electrodes and electrets, small relative displacement can give rise to huge capacitance variation between the top and bottom plate. This characteristic is very suited to MEMS devices with tiny volumes, where the displacement of the seismic mass is constrained within a confined space. Therefore, in-plane gap varying-type harvesters are often employed and investigated in MEMS energy harvesting applications.

Electret materials are usually divided into two categories: polymer-based organic electrets (Teflon, LDPE, PVDF, parylene, CYTOP) and SiO_2-based inorganic electrets (SiO_2, Si_3N_4). One of the critical processes for in-plane gap varying electret harvesters is the micro-patterning of the charge distribution. To obtain stable and high surface potential, micro-sized electrets are the prerequisite step for in-plane electret harvesters. Figure 1.8 shows the summarized micro-patterning charge distribution methods of in-plane gap varying electret harvesters. Figure 1.8a–e shows the micro-patterning polymer-based CYTOP electret by surface chemical etching proposed by Sakane et al. [14]. The interdigital electrodes are first patterned on Pyrex glass by a lift-off process. This is followed by spin-coating CYTOP on its surface and subsequent curing process. After that, the electrodes are further patterned on the

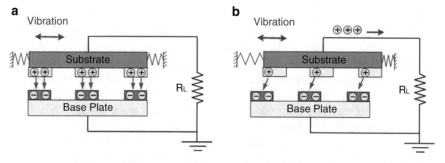

Fig. 1.7 Operating principles of electret-based vibrational energy harvester with in-plane gap varying scheme: (**a**) 100% overlapping, (**b**) with relative displacement

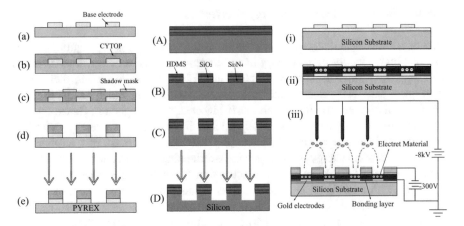

Fig. 1.8 Summarized micro-patterning charge distribution methods: (**a–e**) micro-patterning polymer-based CYTOP electret by surface chemical etching, (**a–d**) micro-patterning silicon-based SiO_2/Si_3N_4 by bulk DRIE etching, (**i–iii**) localized shadow mask patterning without chemical etching process

CYTOP surface serving as a shadow mask. The CYTOP is then chemically patterned and corona charged. Tens of micrometer-sized electret patterned can be obtained with lithographic process; however, the process is rather complicated and costly. The second approach is micro-patterning silicon-based SiO_2/Si_3N_4 by deep reactive-ion etching (DRIE) proposed by Boisseau et al. [15], as shown in Fig. 1.8a–d. Although SiO_2 can achieve a high surface charge density up to 10 mC/m^2, the charge stability is a big problem. Large parasitic capacitance of silicon-based device is also a challenge. Therefore, DRIE process is utilized to create trenches in adjacent electrets. However, the overall output performance is still quite limited.

Figure 1.8i–iii shows a new type of micro-patterning charge distribution by localized corona charging method by Tao et al. [16, 17]. The electret thin film was firstly bonded (insoluble electrets such FEP or LDPE) or spin-coated (soluble electrets such as CYTOP) on the surface of patterned electrodes. A thin silicon wafer is first etched by a wafer-through DRIE process to get the openings. It is then deposited with 300 nm thick gold, which serves as a shadow mask. The electret thin film is selectively charged by being placed between the shadow mask on the top and the glass substrate with gold electrode on the bottom. By this new approach, electret width with tens to hundreds of micrometers can be easily achieved without any chemical etching and removing process, either by surface or bulk micromachining. The whole piece of electret thin films is kept intact during the charging process. It is beneficial to maintain the good charge stability of original electret thin films. Figure 1.9 shows the SEM images of the micro-sized charge distribution on LDPE electret thin film for two-dimensional energy harvesting. The white and black areas depict negatively charged and uncharged areas. It can be seen that very clear micro-sized charge distribution can be achieved.

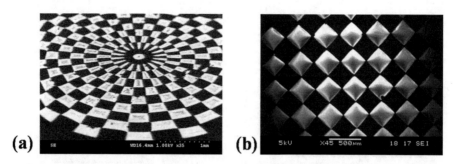

Fig. 1.9 SEM images of the micro-sized charge distribution on LDPE electret thin film for two-dimensional energy harvesting (negatively charged): (**a**) patterned with 200 μm width circular shape [16]. (© IOP Publishing. Reproduced with permission) and (**b**) patterned with 200 μm width square shape [17]. (Elsevier Publishing. Reproduced with permission)

The other type of electret energy harvester works on the out-of-plane gap closing scheme. The capacitance variation depends on the air gap distance change between the two parallel plates. Therefore, neither micro-patterning of electret material nor precise alignment is required, leading to a simple and cost-effective fabrication process. However, stiction induced by electrostatic instability is a problem of concern when the surface potential of electret is too high. The working principle of out-of-plane gap closing scheme is shown in Fig. 1.10. The motion of the top electrode converts the external kinetic energy to electricity by changing the electrostatic induction conditions.

Out-of-plane electret energy harvesters usually adopt a cantilever beam and a seismic mass configuration due to its simple structure and easy formation process. However, it also gives rise to the narrow bandwidth problem where the ambient kinetic energy usually exhibits broad frequency spectrums. To broaden the operating bandwidth, Tao et al. [18, 19] proposed a dual-charged electret vibration system where both negative and positive charged electrets were integrated into single resonant system, as shown in Fig. 1.11a. Figure 1.11b shows the optical image of the fabricated electret energy harvesting device. Elastic polydimethylsiloxane (PDMS) was employed on the surface of the electret to get avoided of "pull-in" effect. The resonant structure is formed by DRIE process in silicon. The bottom plate is formed on a glass substrate. The enhanced electric field by dual-charged electrets has induced a strong electrostatic spring softening nonlinearity in the vibration system. The bandwidth was enlarged from 1.3 Hz of linear response to 3.7 Hz of spring softening response.

To further exploit the energy harvesting capability at two extreme positions of the counter electrode, Tao et al. [20] further investigated a sandwich-structured electret energy harvester that has both positively and negatively charged electret plates located at two extreme positions. The schematic illustration and the 3D structure of the three-plate structure are shown in Fig. 1.12a, respectively. The conventional two-plate electret energy harvester can only induce maximum charge when the movable mass reaches its lowest position. With the newly proposed three-plate structure, the harvester is able to induce maximum charge at both lowest and highest positions,

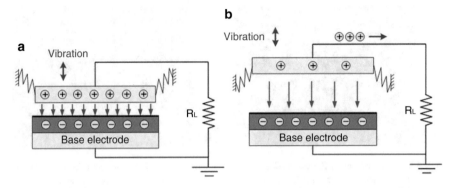

Fig. 1.10 Operating principle of electret-based vibrational energy harvester with out-of-plane gap closing scheme: (**a**) moving close, (**b**) separating apart

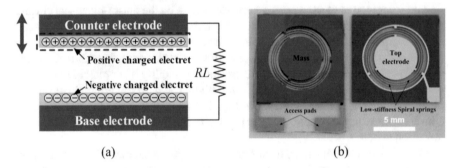

(a)   (b)

Fig. 1.11 (**a**) Schematic of the proposed out-of-plane electret energy harvester with two charged plates, (**b**) optical images of the fabricated device on glass substrate [18]. (© IOP Publishing. Reproduced with permission)

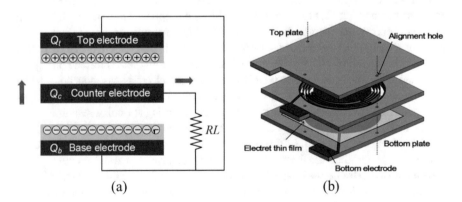

(a)   (b)

Fig. 1.12 (**a**) Schematic illustration of the proposed three-plate sandwich-structured electret energy harvester, (**b**) 3D schematic structure [20]. (© IOP Publishing. Reproduced with permission)

giving rise to higher output performance. The testing results showed the output voltage was enhanced by 80.9% and 18.6% compared to the two-plate configuration with the top electret alone and bottom electret alone configurations, respectively.

Compared with other broadband approaches involving extra mechanical components and tuning efforts, multi-frequency technique exploiting multiple vibration modes of single two-degree-of-freedom (2DOF) system provides a simple and reliable solution to increase the energy harvesting effectiveness. Therefore, Tao et al. [21] developed and investigated a 2DOF e-VEH system that comprised a primary subsystem for power generation and an accessory subsystem for frequency tuning. Figure 1.13 shows a schematic and a photograph of the proposed device structure and the fabricated prototype of the proposed e-VEH device, respectively. By precisely tuning the accessory mass, the first two resonances of primary mass could be tuned close to each other while maintaining comparable magnitudes. With further increased excitation accelerations, it was observed that the 2DOF e-VEH system demonstrated a spring hardening nonlinear effect, where the first peak was capable of being driven toward convergence with the second one to achieve a broadband energy harvesting system. Such novel nonlinear two-degree-of-freedom (2DOF) energy harvesting system combines advantages of both multimodal energy harvesting and the nonlinear technique, which potentially offers new insight of increasing bandwidth with hybrid broadband mechanisms.

Electroactive Polymer-Based Harvesters

There are two types of electroactive polymers (EAPs) that can be used for energy conversion applications, namely, dielectric elastomers (DEs) and ionic polymer-metal composites (IPMCs). Both types of EAPs operate on the electromechanical coupling between the mechanical strain and electrical charge movement.

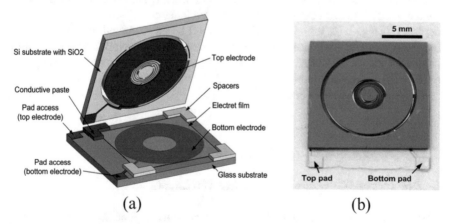

(a)   (b)

Fig. 1.13 (**a**) Schematic and (**b**) fabricated prototype of the proposed electret-based 2DOF MEMS energy harvesting device

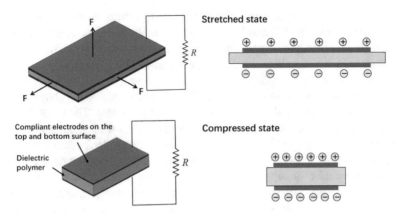

Fig. 1.14 Schematic of basic operational mechanisms of dielectric polymer-based energy harvesters

The electromechanical coupling of DEs mainly relies on the dipole polarization or electrostatic mechanisms, while IPMCs are dominated by the charge diffusion in polymer network.

A typical DE energy harvester is based on a film of an elastically deformable polymer material that is covered with compliant conductive electrodes on both sides, as shown in Fig. 1.14. The working principle of DE-based energy harvester is analogous to that of electrostatic energy conversion. Dielectric elastomer transducers are capable of responding to the applied force and converting the deformation to electricity. A complete energy conversion cycle can be divided into four stages: (a) charging stage, add electrical charge to the polymer while it is in the stretched state; (b) circuit switching stage, switch to the open-circuit condition while the film is relaxed or thickened at the fixed charge stage; (c) discharging stage, connect to the storage circuit and further thickening of the polymer; (d) switching back stage, switch to open-circuit condition with increased tension and reduced thickness. As depicted in Fig. 1.14, an external circuit should be incorporated to providing and removing charges when the dielectric elastomer generator works. The maximum amount of charge generated depends on the charge provided and the material properties, such as the maximum strain imposed, the affordable maximum electric field, and the need to keep the elastic restoring force.

1.2.2   Electromagnetic MEMS Harvesters

Electromagnetic Generators by Relative Movement

Electromagnetic generators have been used to transform mechanical energy to electricity since the early 1930s, such as big wind farms or huge hydropower dams. Over the last one or two decades, electromagnetic generators further evolve into a variety of micro-/nanoscale devices, ranging from cubic micrometers to centimeters. These energy harvesting devices are no longer targeting large-scale power generation from wind farms or dams, but rather power supply for microelectronic devices and WSNs. The basic operating principle of electromagnetic power conversion is based on Faraday's law of induction, where the voltage or electromotive force (emf) can be induced when there is a change of magnetic flux in a conductive loop of wire. The voltage is proportional to the time rate of change of the magnetic flux linkage, as given below

$$V = -\frac{d\Phi}{dt} = NBl_c \frac{dy}{dt} \qquad (1.7)$$

where Φ is the flux linkage in the coil, B is the magnetic field strength, N is the number of turns in the coil, l_c is the length of the turn, and y is the distance the coil moves through the magnetic fields.

The technical and theoretical foundations of energy harvesting through electromagnetic mechanism have already been well established. One of the earliest discussions of micro-electromagnetic energy harvester was presented by William and Yeats in 1996 [22]. They modeled an electromagnetic harvester as a one-degree-of-freedom (1DOF) inertial system, where a movable seismic mass (m) is suspended by a spring (k) and is damped by the electromagnetic damping (η_e) and parasitic mechanical damping (η_m). With a sinusoidal excitation ($y = A \sin(\omega t)$), the output power can be derived as

$$P_{W\&Y} = \frac{m_1 \zeta_t A^2 \omega^3 \left(\dfrac{\omega}{\omega_n}\right)^3}{\left[1 - \left(\dfrac{\omega}{\omega_n}\right)^2\right]^2 + \left[2\zeta_t \left(\dfrac{\omega}{\omega_n}\right)\right]^2} \qquad (1.8)$$

where ω_n is the resonant frequency, ζ_t is the total damping ratio, and A and ω are the amplitude and frequency of the excitation, respectively. Maximum power can be achieved when electromagnetic damping is equal to the mechanical parasitic damping. The electrical output power is 50% of the total power from the frame vibration.

Rare earth permanent magnets, such as NdFeB and SmCo, exhibit the best performance and are widely used in various electromagnetic devices. One of the greatest challenges hindering the development of micro-electromagnetic generator is the formation of the rare earth permanent magnet in MEMS fabrication process. In traditional magnetic devices, the magnets are usually first manufactured and then assembled into electromechanical system to perform some specific functions, such as power generation or magnetic actuation. However, the MEMS process requires monolithic integration of the magnets within a multistep sequential process flow. Therefore, Tao et al. [23, 24] have proposed a fully integrated micro-electromagnetic

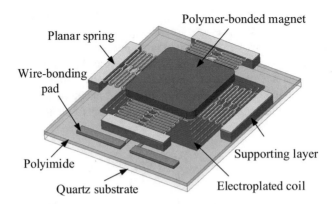

Fig. 1.15 3D Schematic view of the proposed fully integrated magnetic harvester with micro-patterned magnets. (© Elsevier Publishing. Reproduced with permission)

vibration energy harvester with micro-patterning of bonded magnets. The 3D schematic is shown in Fig. 1.15. Components such as the induction coil, the suspended planar spring, the supporting layer, and the wire-bonding pad can be batch-fabricated using layer-by-layer sacrificial surface micromachining technology.

The bonded magnet was made from commercially available NdFeB powder which was dispersed in room temperature-cured epoxy resin. The entire micro-electromagnetic generator is fabricated by MEMS-compatible laminated surface micromachining process, which is devoid of any assembly procedure. The detailed fabrication flow and fabricated device are shown in Figs. 1.16 and 1.17, respectively. It involves planar coil sputtering, polyimide spin coating, supporting layer electroplating, bonded magnet micro-patterning, and final release process. This shows the possibility to integrate an array of micro energy harvesters through a single batch-fabricated process flow. The size of the entire structure is $4.5 \times 4.5 \times 1$ mm^3, which is among some of the smallest electromagnetic vibration energy harvesters demonstrated up to date.

Two-degree-of-freedom (2DOF) energy harvesters have been widely utilized in macro piezoelectric energy harvesters. However, 2DOF MEMS devices with electromagnetic mechanisms have rarely been exploited. The main difficulty is the precise control of the mass ratio (μ) and frequency tuning ratio (α) in order to achieve two close and comparable peaks. Tao et al. [25] have proposed a 2DOF electromagnetic energy harvesting (EM-EH) chip that comprises mainly of a 2DOF spring-mass resonators fabricated with a wafer-level batch fabrication process on SOI wafers. Induction coil is only patterned on the primary mass for energy conversion. The 3D schematic and fabricated EM-EH chip of proposed 2DOF EM-EH device is shown in Fig. 1.18a, b, respectively. The detailed fabrication process of proposed 2DOF EM-EH chip on SOI wafer is further depicted in Fig. 1.19. In this work a 2DOF MEMS energy harvesting device is demonstrated where two close resonances with comparable magnitudes can be achieved through the precise control of circular beam-mass system by the merit of MEMS fabrication process.

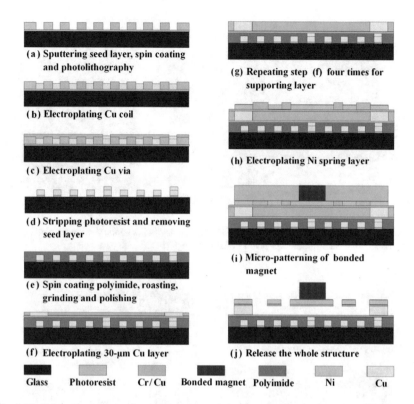

(a) Sputtering seed layer, spin coating and photolithography

(b) Electroplating Cu coil

(c) Electroplating Cu via

(d) Stripping photoresist and removing seed layer

(e) Spin coating polyimide, roasting, grinding and polishing

(f) Electroplating 30-μm Cu layer

(g) Repeating step (f) four times for supporting layer

(h) Electroplating Ni spring layer

(i) Micro-patterning of bonded magnet

(j) Release the whole structure

Glass   Photoresist   Cr/Cu   Bonded magnet   Polyimide   Ni   Cu

Fig. 1.16 Fabrication process flow of the proposed electromagnetic power generator. (© Elsevier Publishing. Reproduced with permission)

Fig. 1.17 Optical photographs of the fabricated energy harvester prototypes: (**a**) device configuration, (**b, c**) optical and SEM image of planar coils by electroplating, (**d**) harvester arrays, (**e**) planar coil array embedded in polyamide, (**f, g**) two types of released Ni supporting springs. (© Elsevier Publishing. Reproduced with permission)

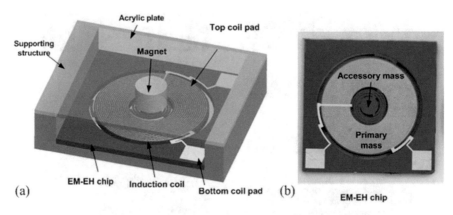

Fig. 1.18 (**a**) 3D schematic of proposed 2DOF EM-EH device; (**b**) optical image of the fabricated spring-mass resonant structure. (© IOP Publishing. Reproduced with permission)

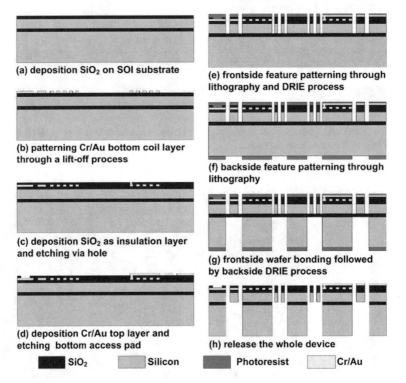

Fig. 1.19 Fabrication process of proposed 2DOF EM-EH chip on SOI wafer. (© IOP Publishing. Reproduced with permission)

Electromagnetic Generators with Magnetostrictive Materials

Typically, Faraday's law of electromagnetic induction is usually implemented by the relative displacement variation between the magnet and the coils. Some special types of magnetostrictive materials are capable of converting kinetic strain/stress to magnetic field/induction variation, known as *Villari effect*. If the magnetic field change is captured by an induction coil, the mechanical strain can also be converted to electricity. This has some similarity with the piezoelectric material, where the mechanical strain can be transformed to the charge circulation. Therefore, these materials can also be called as piezomagnetic materials. There are several types of magnetostrictive materials such as Terfenol (terbiumdisprosium-iron alloy), Galfenol (gallium-iron alloy), and Alfenol (aluminum-iron alloy). Galfenol is one of the most common magnetostrictive materials used in the energy harvesting applications due to its high piezomagnetic constant. More than 1 T change of magnetic induction can be achieved with stress variations. The magnetostrictive vibrational energy harvester operates on two principles: force-driven harvesters and vibration-based harvesters [26].

One of such magnetostrictive harvesters was developed by Ueno et al. [27] based on Galfenol alloy wound with coil and York structure. The permanent magnet is employed on the back of the York to provide a magnetic bias and facilitate to form a closed loop of magnetic flux. One end of the two-parallel magnetostrictive Galfenol is fixed to the handle by soldering. The other end is connected to the seismic mass. When there is an external excitation, the oscillation of the seismic mass on the tip of the lever will continuously press and stretch the magnetostrictive rod. The variation of the magnetization generated will give rise to a current in the winding coil. With a miniaturized Galfenol alloy rod of $2 \times 0.5 \times 7$ mm^3, it is reported that an average power density of 3 mW/cm^3 could be achieved at 212 Hz at an excitation of 1.2 g.

1.2.3   Piezoelectric MEMS Harvesters

Piezoelectric materials possess the unique merit of direct electromechanical coupling that can convert mechanical strain to electrical energy and vice versa. Piezoelectric energy harvesters make use of external excitations to continuously compress or stretch piezoelectric materials to convert kinetic energy to electrical energy. The piezoelectric energy harvester mainly works on two principles, directly excited by external force or by an inertial mass oscillation. The performance of the piezoelectric energy harvesters is highly dependent on the piezoelectric properties of the materials. Generally, the piezoelectric materials can be categorized into several forms, including piezoceramic (PZT or lead zirconate titanate), barium titanate (BaTiO$_3$), single crystal (quartz), thin film (ZnO or AlN), thick film based on piezoceramic powder, and polymeric materials (PVDF). Currently, piezoceramics, such as PZT and micro-fiber composites (MFC), are the most commonly used piezoelectric materials in energy harvesting applications due to their high piezoelectric coefficient. PVDF thin film is another widely used as flow or acoustic sensors due to its low cost and high flexibility, but its piezoelectric coefficient is comparatively low.

Most piezoelectric materials for energy harvesters have well-defined polar axis and exhibit anisotropic characteristics. The piezoelectric strain coefficient has two subscripts which indicate the directions of the polar axis and the strain. The polar axis is referred to "3" direction. Other directions at right angles to the polar axis are referred to "1" direction symmetrically. The applied strain to the piezoelectric materials can be either in the same direction along or perpendicular to the polarization axis. Therefore, there are two most common modes used in piezoelectric energy harvesting [9]. A d_{33} mode piezoelectric energy harvester indicates the applied stress/strain is parallel to the polar axis and the generated power is along the same axis, as shown in Fig. 1.20a. The d_{31} mode indicates the applied stress/strain is at the right angle to the polar axis and the generated power is perpendicular to the stress, as shown in Fig. 1.20b.

For ferroelectric ceramics (PZT or MFC) or polymer (PVDF), the polar direction is defined during the post-polarization or annealing process. A few other piezoelectric thin films, such as AlN and ZnO, are nonferroelectric crystalline materials. The polar axis is dominated by the crystal orientation during the deposition process. Generally, both piezoelectric coefficient and coupling constant of d_{33} mode are higher than that of d_{31} mode. However, the operation of d_{33} mode can achieve large strain in "1" direction, since d_{31} mode has comparative low Young's modulus. Therefore, d_{31} mode is more commonly used in energy harvesting applications. Lee et al. fabricated piezoelectric energy harvesters with two different operation modes, d_{31} mode and d_{33} mode [28]. To achieve high piezoelectric coupling of d_{33} mode, interdigital electrodes have been deposited on the surface of the piezoelectric thin films.

1.2.4   Triboelectric Nanogenerators

Triboelectric nanogenerators are first proposed by Z. L. Wang's group. Triboelectric nanogenerators work on the combined effect of contact triboelectrification and electrostatic induction. Contact triboelectrification is the phenomenon of charge transfer that takes place when two dissimilar materials are put into contact. Electrons can be

Fig. 1.20 Piezoelectric energy harvesters with two operating types: (**a**) d_{31} mode and (**b**) d_{33} mode

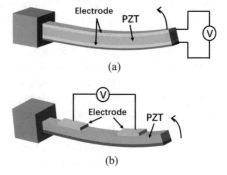

(a)

(b)

transferred between two contact materials due to the different work functions of the contacting materials, either in metal to insulator or insulator to insulator [29]. Depending on the electron trap tendency of various materials during the triboelectrification process, the materials can be arranged in a series, as summarized in Fig. 1.21. The materials arranged on the positive side are more likely to donate electrons, while the negative-side materials are inclined to accept electrons. Therefore, appropriate materials should be selected when the two dissimilar materials come into contact with each other. Generally, the materials with farther relative positions can generate more power during the triboelectrification process.

Typically, on the basis of operating mechanisms, the triboelectric nanogenerators mainly work on two mechanisms, including out-of-plane vertical contact-separation scheme and in-plane lateral sliding scheme. The overall principle is based on electrostatic induction, which is similar to the electret-based electrostatic energy harvester (e-VEH). The most notable difference is that the charge generated from triboelectric effect is due to the contact electrification, while the charge in e-VEH is by pre-implanting process. Specifically, for out-of-plane vertical contact-separation scheme (Fig. 1.22), the charge is created when the two dielectrics with different electron affinities contact with each other (Fig. 1.22b). The charge on the dielectrics would induce opposite charge in the back electrodes (Fig. 1.22c). The capacitance change would lead to the induced charge that flows back and forth between the two electrodes (Fig. 1.22a–d).

The in-plane lateral sliding operation scheme relies on the variance of the overlapping area of two dielectrics, as shown in Fig. 1.23. As the triboelectric layers get in contact, there is equal charge generated between the metal electrodes and the polymer. It should be mentioned that the charge located in metal electrodes is dependent on the electric field distribution. As the sliding triboelectric layer moves apart (Fig. 1.23b), there is a current flow from the top electrode to bottom electrode due to the electric field of polymer difference. The flow continues to the maximum state until the two layers are completely separated (Fig. 1.23c). The charge moves back and forth when the two layers continue sliding, transforming mechanical energy to electrical energy. Triboelectric nanogenerators are one of the most popular energy harvesters that are intensively investigated to serve as self-powered sensors in various applications.

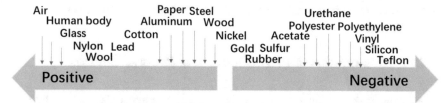

Fig. 1.21 Materials in triboelectric generator with different tendency to attract and donate electrons

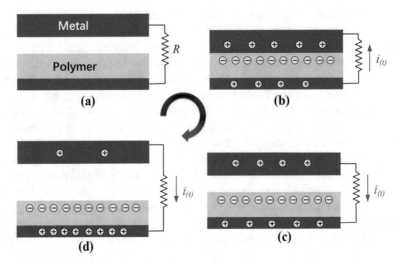

Fig. 1.22 Schematic for out-of-plane vertical contact-separation scheme-based triboelectric nano-generator with polymer and metal contact: (**a**) original state, (**b**) in contact, (**c**) splitting apart, (**d**) at the maximum separation

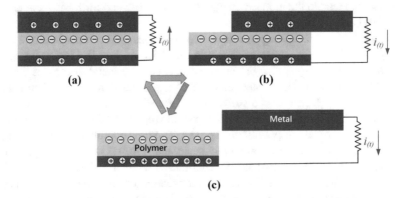

Fig. 1.23 Schematic for in-plane lateral sliding scheme-based triboelectric nanogenerator with polymer and metal contact: (**a**) in contact, (**b**) sliding apart, (**c**) at the maximum separation

1.3   MEMS/NEMS-Enabled Harvesters as Self-Powered Sensors

1.3.1   Ceramic-Based Piezoelectric Self-Powered Sensors

The piezoelectric coefficients of ceramic piezoelectric materials, such as PZT or $BaTiO_3$, are an order of magnitude higher than that of nonferroelectric materials (ZnO or AlN). Therefore, ceramic piezoelectric materials are more advantageous in developing self-powered sensors in terms of low signal-to-noise ratio and absolute output performance. One of the interesting works is done by Liu et al. [30]. They have developed a PZT-based piezoelectric microcantilever which can function as both an airflow sensor and a wind-driven energy harvester. Compared to the conventional micro-flow sensors that rely on resistive-strain change or thermal flux variation, the proposed PZT-based piezoelectric sensor does not require external power sources. In addition, the wind-driven energy harvesting gives rise to the oscillations of the harvester at its natural frequency. Therefore, it can eliminate the bandwidth issue of conventional vibration energy harvesters by combining the sensing element and aerodynamic power generating function together. The PZT microcantilever is fabricated through a laminating MEMS process and features a length of 3000 μm and a width of 300 μm. Through this approach, the flow sensing sensitivity up to 0.9 mV/(m/s) is achieved. Another similar self-powered flow sensor is developed by Hu et al. which is based on traditional cantilever mass piezoelectric harvesters [31]. They fabricated cantilever-based micromachined MEMS energy harvesters using self-powered wind sensors. Because it has bulky cantilever mass system with comparative large volume, the critical wind speed of the device is between 12.7 and 13.2 m/s. Therefore, it can be regarded more as a MEMS airflow energy harvester rather than a MEMS wind sensor. Besides thin-film ceramic piezoelectric materials, crystalline PZT nanoribbons could achieve even higher piezoelectric coefficient due to the nanoscale features [32]. The PZT nanoribbons could be formed on the PDMS substrate. With a transfer printing method, the PZT ribbons were first deposited on a pre-strained PDMS substrate and then spontaneously formed with wavy/buckled pattern when PDMS was released.

1.3.2   Self-Powered Sensors Using ZnO Nanowires

ZnO nanowires, as one-dimensional nanomaterials, have both excellent semiconducting and piezoelectric properties with large energy band gap (3.37 eV) and excitation binding energy (60 meV), forming the basis of electromechanical coupling sensors and transducers. The fundamental mechanism is that when the ZnO nanowire is strained by external force, a piezo potential is generated between the nanowire and the counter contact (metal tips) to balance the Fermi level difference, driving a transient flow of electrons in the external load circuit. The piezoelectric nanogenerator based on ZnO nanowire array was first proposed by Z. L. Wang in 2006 [33]. An output voltage of 8 mV was detected when AFM tip was employed to scan one of the nanowires. In 2008, Wang et al. [34] proposed the density-controlled growth of aligned ZnO nanowires without using ZnO seeds. By this method, the ZnO nanowires can be grown in a large scale on different types of substrates. One of the interesting researches conducted by Z. L. Wang's group is to integrate such large-scale ZnO nanowire arrays into the rotating tire of mobile vehicles [35]. The ZnO nanowires were planted on a polyester (PS) substrate. The nanowire textured thin film was adhered tightly to the inner surface of the tire. Electrical pulse could be generated each time when the nanowires were bent. These nanowires could be used to develop as self-powered speed and tire pressure sensors. The harvested kinetic energy could be further used to power up a liquid crystal display (LCD) screen. Lee et al. [36] developed another type of self-powered chemical sensor based on ZnO nanowire and single-walled CNT (SWNT) transistor for Hg^+ concentration detection. When there were mercury ions existed in water, a LED indicator can be light up automatically. It was also the first demonstration of using self-powered sensor system for detecting toxic pollutant in the environment.

1.3.3 Polymer-Based Piezoelectric Nanosensors

Polymeric polyvinylidene fluoride (PVDF) nanofibres feature a unique combination of natural flexibility and stretchability as well as excellent piezoelectric coefficient, making it as a promising candidate for wearable energy harvesting applications and flexible sensors. In nature, PVDF polymer consists of several different types of structural forms depending on chain conformation of hydrogen and fluoride sequence along with the backbone. Naturally, PVDF polymer is mostly available in α-phase when it is cooled and solidified from melting state. However, only β-phase has piezoelectric response due to the aligned hydrogen and fluoride cells. β-phase usually can be achieved by electrical poling as well as stretching process. Through electrospinning process, PVDF melts or solutions can be formed in fibers with diameters ranging from nanometer to micrometer scale. Chang et al. [37] successfully fabricated PVDF nanofibre-based nanogenerators through near-field electrospinning process combined with direct-write mechanical stretching process. It can be used to detect cycling strain variation within a flexible substrate. The output current of 5 V could be achieved in stretch and release under cycled strain at 3 Hz.

1.4 Triboelectric Nanogenerators as Self-Powered Sensors

1.4.1 Triboelectric Nanogenerators as Trajectory Sensors

Recent advances in integrated circuit (IC) manufacturing techniques and Internet of Things (IoT) lay demands on the availability of microelectronic systems that are wireless, portable, multifunctional, and easy to operate. The development of self-powered sensors making use of ambient energy is highly desirable, considering a large amount of wireless sensor nodes (WSNs) to be developed for various applications ranging from national defenses, health and environment monitoring, smart sensing, and telecommunications to medical implants. Triboelectric nanogenerators are able to convert mechanical energy such as motion and force to electrical energy by triboelectrification and electrostatic induction. Alternatively, the electrical information created from nanogenerators, such as voltage, current, and power, can also be used to predict the mechanical input parameters such as magnitude, frequency, and force.

Triboelectric nanogenerators can operate on different types of motions such as linear sliding, rolling, and rotation. Since the electrical information is directly related to these motion characteristics, triboelectric nanogenerators can be utilized to track the trajectory and detect real-time parameters of specific motions. For sliding motions, Zhou et al. [38] proposed a single-axis displacement and speed sensors with a pair of micro-grating layers with interdigital electrode patterns. The bottom layer was fabricated from silicon oxide wafer with Al electrode to generate positive charge. The top counter electrode was formed from glass substrate coated with ITO and parylene. Within the sensitive region from 10 to 190 μm, the resolution of each step with 1 μm can be clearly identified. The corresponding output voltage variation of each step is 2.2 mV with a noise of 0.38 mV. In addition, rotation is another common motion form widely existing in turbines, gears, or automobiles, etc. Lin et al. [39] developed a disk-type rotational nanogenerators with segmental structures, which could serve as self-powered angular speed sensor for rotational motions. The rotation speed of the disk can be dominated by both the frequencies of the magnitude of the output power. On further improvement [40], both the radial and axial movement of a braking pad can be detected simultaneously with the free-rotating disk nanogenerators. Overall, compared to other types of sensors, triboelectric self-powered sensors have the unique advantages in terms of broad material selection and excellent adaptability.

1.4.2   Triboelectric Nanogenerators as Tactile Sensors

Since the output performance of triboelectric nanogenerator is highly influenced by external mechanical stimuli, the most straightforward approach is to develop self-powered tactile sensors. One of the pioneering researches is done by Fan et al. [41]. They proposed a triboelectric-based tactile sensor with micro-patterned plastic surface on a transparent substrate. The contact triboelectric effect within the two dissimilar polymer sheets could respond to very gentle pressure change, as light as a single water drop or a falling feather. The applied pressure and the output potential difference were found in a linear relationship, indicating a suitable sensing method. The output voltage could precisely predict the static external pressure, while the current density was able to measure the dynamic load conditions. Long et al. [42] further integrated a 6 × 6 pressure sensor array onto an aluminum sheet for dynamic pressure mapping. A contour plot of the pressure mapping could be visualized by the output voltage of each triboelectric nanogenerators. One of the main limitations is that the area of each unit is up to 2×2 cm^2, making the spatial resolution of tactile imaging quite low. However, it opens up the new application potential in self-powered touchscreen and smart skins.

Zhu et al. [43] further developed flexible triboelectric tactile sensors with PET/ITO/FEP/nylon multilayer structure. FEP thin films were patterned with nanowires to assist the triboelectric charges generation. Depending on the location region of the polymer, the pressure sensitivity is ranging from 0.5 mV/Pa to 44 mV/Pa. Practical applications were further demonstrated by the self-powered tactile sensors, including wireless siren alarm system on the door handle and lighting system upon contact of human fingers. Since the material selection is broad, the triboelectric tactile sensors have huge potential applications, such as healthcare and security.

1.4.3   Triboelectric Nanogenerators as Acoustic Sensors

Although the power density of acoustic energy is comparatively low, the acoustic wave ubiquitously exists in the ambient environment. Therefore, developing self-powered acoustic sensors has also attracted much attention. From energy harvesting perspective, Helmholtz resonator is one of the most popular methods to convert acoustic wave vibration to electric energy by using second-order fluidic oscillation. Helmholtz resonance is the phenomenon of air resonance in a gas-filled cavity with an open neck. It can be simply represented as a spring-mass system, where the air in the neck acts as the mass and the air inside the cavity represents the spring. The Helmholtz resonant frequency depends on the volume of the cavity and the neck. Yang et al. [44] invented the first self-powered acoustic sensor with triboelectric-based Helmholtz resonators. They replaced the wall of cavity with nanowire-patterned PTFE thin film and nanopore-patterned aluminum. The triboelectrification would take place when the acoustic wave was incident on the two layer surfaces. Sound frequency bandwidths ranging from 10 to 1700 Hz could be recorded with different dimensions of the triboelectric generators. A self-powered thin-film microphone for sound recording which uses this type of triboelectric generators was also demonstrated. This type of self-powered acoustic sensors can also be used in three-dimensional underwater acoustic source positioning [45]. By mimicking the human ear functions, high sensitivity and good adaptability could be achieved in the underwater environment. The application can be further broadened to military surveillance monitoring and wireless sensing applications in the future.

1.4.4   Triboelectric Nanogenerators as Chemical Sensors

The output characteristics of triboelectric nanogenerators not only rely on the external mechanical excitations (magnitude, frequency, and force) but also are strongly affected by the material types and contact surface modifications. Therefore, through analyzing the signal outputs, certain triboelectric nanogenerators can be used as self-powered chemical sensors to detect the special ion concentration, heat, humidity, UV illumination, etc. One of the pioneering works is conducted by Lin et al. [46] for Hg^{2+} detection. They employed Au nanoparticles with 3-mercaptopropionic acid (3-MPA molecules) in the typical contact-mode triboelectric nanogenerators. The surface polarity of the metal plate would undergo changes subjected to chemical bond between Hg^{2+} ions and 3-MPA molecules. Therefore, Hg^{2+} ion concentration could be real-time monitored by the output performance of triboelectric nanogenerators. It was found that the output current density was in a linear relationship with the Hg^{2+} ion concentration. The detection limit of Hg^{2+} ion concentration was as low as 10 nM to 5 mM. Another work done by Li et al. [47] demonstrated the phenol detection and electrochemical degradation by self-powered triboelectric chemical sensors. They used β-cyclodextrin (β-CD) both as triboelectrification enhancement element and phenol molecule recognition detector in the chemical solution. They proved that the decrease of output current to phenol was much larger than that of other controlled organic species, indicating good selectivity of the proposed self-powered sensors. The advancement of the proposed self-powered sensor can be further used in environmental monitoring, treatment, degradation, and sustainability.

1.5 Conclusions

Recent surge in Internet of Things (IoT) and micro-/nanotechnologies shifts the research focus from single-functioned discrete devices to more intelligent, multifunctional, and self-sustainable microsystems, including sensing and actuating, communicating, controlling, and self-powering functions. The development of self-powered active sensors by capturing and transforming unused ambient energy into the electrical energy would be the optimum solution. It has the best adaptability and does not require the complicated energy management and coupling between the power source and sensing element.

The advancement of micro-/nanoelectromechanical systems (MEMS/NEMS)-enabled energy harvesting technologies in the last decade has yield breakthroughs in self-powered sensor evolutions. This chapter reviews the latest advancements of MEMS/NEMS-enabled energy harvesters as self-powered sensors. Several types of vibration-to-electricity conversion mechanisms are intensively reviewed and compared, including electrostatic, electromagnetic/magnetostrictive, piezoelectric, and triboelectric. There is no doubt that the self-powered sensors would play a critical role in the next-generation sensor network deployment and vast intelligent MEMS/NEMS systems.

Even though a significant progress has been made in the last decade in terms of both technological improvement and fundamental understanding, MEMS/NEMS-enabled energy harvesters as self-powered sensors are still an emerging technology, and the full potential of such devices has not been intensively explored yet. Several issues should be appropriately addressed before the self-powered sensors can be widely adapted. One critical challenge is the compatible fabrication with batch microfabrication process for mass production. However, it is believed that MEMS/ NEMS-enabled self-powered sensor technology is the primary step forward to enable next generation of wireless sensing and portable electronic applications. The large deployment of such smart sensors is expected to have revolutionary influence in future infrastructure/environmental monitoring, healthcare, smart homes/cities, etc. We expect that more researchers will get involved in this exciting field in the next decade to make our life more convenient and wonderful.

Acknowledgments This research is supported by National Natural Science Foundation of China (Grant No. 51705429), the Fundamental Research Funds for the Central Universities, and the Key Laboratory fund of Science and Technology on Micro-system Laboratory (No. 614280401010417).

References

1. Sauerbrey, J., Schmitt-Landsiedel, D., & Thewes, R. (2003). A 0.5-V 1-/spl mu/W successive approximation ADC. *IEEE Journal of Solid-State Circuits, 38*(7), 1261–1265.
2. Steingart, D. (2009). Power sources for wireless sensor networks. In S. Priya & D. Inman (Eds.), *Energy harvesting technologies* (pp. 267–286). New York: Springer.
3. Knight, C., Davidson, J., & Behrens, S. (2008). Energy options for wireless sensor nodes. *Sensors, 8*(12), 8037–8066.
4. Paradiso, J. A., & Starner, T. (2005). Energy scavenging for mobile and wireless electronics. *IEEE Pervasive Computing, 4*(1), 18–27.
5. Xie, J., Chengkuo, L., & Hanhua, F. (2010). Design, fabrication, and characterization of CMOS MEMS-based thermoelectric power generators. *Journal of Microelectromechanical Systems, 19*(2), 317–324.
6. Roundy, S., Wright, P. K., & Rabaey, J. (2003). A study of low level vibrations as a power source for wireless sensor nodes. *Computer Communications, 26*(11), 1131–1144.
7. Hudak, N. S., & Amatucci, G. G. (2008). Small-scale energy harvesting through thermo-electric, vibration, and radiofrequency power conversion. *Journal of Applied Physics, 103*, 101301.
8. Zhou, S. X., & Zuo, L. (2018). Nonlinear dynamic analysis of asymmetric tristable energy harvesters for enhanced energy harvesting. *Communications in Nonlinear Science and Numerical Simulation, 61*, 271–284.
9. Chen, G. J., Li, Y. F., Xiao, H. M., & Zhu, X. (2017). A micro-oscillation-driven energy harvester based on a flexible bipolar electret membrane with high output power. *Journal of Materials Chemistry A, 5*, 4150–4155.
10. Halim, M. A., et al. (2018). An electromagnetic rotational energy harvester using sprung eccentric rotor, driven by pseudo-walking motion. *Applied Energy, 217*, 66–74.
11. Zhang, X., et al. (2018). Broad bandwidth vibration energy harvester based on thermally stable wavy fluorinated ethylene propylene electret films with negative charges. *Journal of Micromechanics and Microengineering, 28*, 065012.
12. Wang, Z. L. (2013). Triboelectric nanogenerators as new energy technology for self-powered systems and as active mechanical and chemical sensors. *ACS Nano, 7*(11), 9533–9557.
13. Roundy, S., Wright, P. K., & Pister, K. S. (2002). Micro-electrostatic vibration-to-electricity converters. *Fuel Cells (methanol), 220*(22), 1–10.
14. Sakane, Y., Suzuki, Y., & Kasagi, N. (Oct 2008). The development of a high-performance perfluorinated polymer electret and its application to micro power generation. *Journal of Micromechanics and Microengineering, 18*(10), 104011.

15. Boisseau, S., Duret, A.-B., Chaillout, J.-J., & Despesse, G. (2012). New DRIE-patterned electrets for vibration energy harvesting. In *EPJ Web of Conferences* (Vol. 33, p. 02010). EDP Sciences. Les Ulis, France.
16. Tao, K., Liu, S., Lye, S. W., Miao, J., & Hu, X. (2014). A three-dimensional electret-based micro power generator for low-level ambient vibrational energy harvesting. *Journal of Micromechanics and Microengineering, 24*(6), 065022.
17. Tao, K., Miao, J., Lye, S. W., & Hu, X. (2015). Sandwich-structured two-dimensional MEMS electret power generator for low-level ambient vibrational energy harvesting. *Sensors and Actuators A: Physical, 228*, 95–103.
18. Tao, K., Lye, S. W., Miao, J., Tang, L., & Hu, X. (2015). Out-of-plane electret-based MEMS energy harvester with the combined nonlinear effect from electrostatic force and a mechanical elastic stopper. *Journal of Micromechanics and Microengineering, 25*(10), 104014.
19. Tao, K., Lye, S. W., Miao, J., & Hu, X. (2015). Design and implementation of an out-of-plane electrostatic vibration energy harvester with dual-charged electret plates. *Microelectronic Engineering, 135*(0), 32–37.
20. Tao, K., Wu, J., Tang, L., Hu, L., Lye, S. W., & Miao, J. (2017). Enhanced electrostatic vibrational energy harvesting using integrated opposite-charged electrets. *Journal of Micromechanics and Microengineering, 27*(4), 044002.
21. Tao, K., Tang, L. H., Wu, J., Lye, S. W., Chang, H. L., & Miao, J. M. (2018). Investigation of multimodal electret-based MEMS energy harvester with impact-induced nonlinearity. *Journal of Microelectromechanical Systems, 27*(2), 276–288.
22. Williams, C. B., & Yates, R. B. (Mar-Apr 1996). Analysis of a micro-electric generator for microsystems. *Sensors and Actuators a-Physical, 52*(1–3), 8–11.
23. Tao, K., Ding, G., Wang, P., Yang, Z., & Wang, Y. (2012). Fully integrated micro electromagnetic vibration energy harvesters with micro-patterning of bonded magnets. *Micro Electro Mechanical Systems (MEMS), 2012 IEEE 25th International Conference on*, 2012, pp. 1237–1240.
24. Tao, K., Wu, J., Kottapalli, A. G. P., et al. (2017). Micro-patterning of resin-bonded NdFeB magnet for a fully integrated electromagnetic actuator. *Solid-State Electronics, 138*, 66–72.
25. Tao, K., Wu, J., Tang, L., et al. (2016). A novel two-degree-of-freedom MEMS electromagnetic vibration energy harvester. *Journal of Micromechanics and Microengineering, 26*(3), 035020.
26. Davino, D. Kinetic energy harvesting by magnetostrictive materials. Available: http://www.sigmaaldrich.com/technical-documents/articles/materials-science/kinetic-energy-harvesting.html
27. Ueno, T. (2015). Performance of improved magnetostrictive vibrational power generator, simple and high power output for practical applications. *Journal of Applied Physics, 117*, 17A740.
28. Lee, B., Lin, S., Wu, W., Wang, X., Chang, P., & Lee, C. (2009). Piezoelectric MEMS generators fabricated with an aerosol deposition PZT thin film. *Journal of Micromechanics and Microengineering, 19*(6), 065014.
29. Wang, P. H., et al. (2018). Complementary electromagnetic-triboelectric active Sensor for detecting multiple mechanical triggering. *Advanced Functional Materials, 1705808*, 1–9.
30. Liu, H., Zhang, S., Kathiresan, R., Kobayashi, T., & Lee, C. (2012). Development of piezoelectric microcantilever flow sensor with wind-driven energy harvesting capability. *Applied Physics Letters, 100*(22), 223905–223903.
31. Xuefeng, H., Zhengguo, S., Yaoqing, C., & You, Z. (2013). A micromachined low-frequency piezoelectric harvester for vibration and wind energy scavenging. *Journal of Micromechanics and Microengineering, 23*(12), 125009.
32. Qi, Y., Kim, J., Nguyen, T. D., Lisko, B., Purohit, P. K., & McAlpine, M. C. (2011). Enhanced piezoelectricity and stretchability in energy harvesting devices fabricated from buckled PZT ribbons. *Nano Letters, 11*(3), 1331–1336.
33. Wang, Z. L., & Song, J. (2006). Piezoelectric nanogenerators based on zinc oxide nanowire arrays. *Science, 312*(5771), 242–246.

34. Xu, S., Lao, C., Weintraub, B., & Wang, Z. L. (2008). Density-controlled growth of aligned ZnO nanowire arrays by seedless chemical approach on smooth surfaces. *Journal of Materials Research, 23*(8), 2072–2077.
35. Hu, Y., Xu, C., Zhang, Y., Lin, L., Snyder, R. L., & Wang, Z. L. (2011). A nanogenerator for energy harvesting from a rotating tire and its application as a self-powered pressure/speed sensor. *Advanced Materials, 23*(35), 4068–4071.
36. Lee, M., Bae, J., Lee, J., Lee, C.-S., Hong, S., & Wang, Z. L. (2011). Self-powered environmental sensor system driven by nanogenerators. *Energy & Environmental Science, 4*(9), 3359–3363.
37. Chang, C., Tran, V. H., Wang, J., Fuh, Y.-K., & Lin, L. (2010). Direct-write piezoelectric polymeric nanogenerator with high energy conversion efficiency. *Nano Letters, 10*(2), 726–731.
38. Zhou, Y. S., et al. (2014). Nanometer resolution self-powered static and dynamic motion sensor based on micro-grated triboelectrification. *Advanced Materials, 26*(11), 1719–1724.
39. Lin, L., et al. (2013). Segmentally structured disk triboelectric nanogenerator for harvesting rotational mechanical energy. *Nano Letters, 13*(6), 2916–2923.
40. Lin, L., Wang, S., Niu, S., Liu, C., Xie, Y., & Wang, Z. L. (2014). Noncontact free-rotating disk triboelectric nanogenerator as a sustainable energy harvester and self-powered mechanical sensor. *ACS Applied Materials & Interfaces, 6*(4), 3031–3038.
41. Fan, F.-R., Lin, L., Zhu, G., Wu, W., Zhang, R., & Wang, Z. L. (2012). Transparent triboelectric nanogenerators and self-powered pressure sensors based on micropatterned plastic films. *Nano Letters, 12*(6), 3109–3114.
42. Lin, L., et al. (2013). Triboelectric active sensor array for self-powered static and dynamic pressure detection and tactile imaging. *ACS Nano, 7*(9), 8266–8274.
43. Zhu, G., et al. (2014). Self-powered, ultrasensitive, flexible tactile sensors based on contact electrification. *Nano Letters, 14*(6), 3208–3213.
44. Yang, J., Chen, J., Liu, Y., Yang, W., Su, Y., & Wang, Z. L. (2014). Triboelectrification-based organic film nanogenerator for acoustic energy harvesting and self-powered active acoustic sensing. *ACS Nano, 8*(3), 2649–2657.
45. Yu, A., et al. (2015). Self-powered acoustic source locator in underwater environment based on organic film triboelectric nanogenerator. *Nano Research, 8*(3), 765–773.
46. Lin, Z. H., et al. (2013). A self-powered triboelectric nanosensor for mercury ion detection. *Angewandte Chemie International Edition, 52*(19), 5065–5069.
47. Li, Z., et al. (2015). β-cyclodextrin enhanced triboelectrification for self-powered phenol detection and electrochemical degradation. *Energy & Environmental Science, 8*(3), 887–896.

Chapter 2
Flexible and Wearable Piezoelectric Nanogenerators

Debarun Sengupta and Ajay Giri Prakash Kottapalli

In this age of advanced smartphones and wearable devices, the need for unlimited power has become a basic necessity. Most of the gadgets rely on some sort of power source in the form of batteries or power adapters. For example, smartwatches have become very common these days and have a huge potential for implementation of energy harvesters. In the near future, it will be highly desirable to have self-powered smart wearable devices which meet their energy needs by scavenging mechanical energy produced by physical activities. In order to solve the problem of fast battery depletion in modern smart devices, a significant amount of research has been carried out in the field of energy harvesters especially using thin film technologies and polymer nanofibers. Nanogenerators using polymers with piezoelectric properties like PVDF have attracted a special attention due to their low production cost and high conversion efficiency. Polymer-based nanofiber energy harvesters are not only relevant for wearable devices and smartphones but also for biomedical energy scavenging applications primarily due to their biocompatibility. This chapter particularly deals

D. Sengupta
Department of Advanced Production Engineering, Engineering and Technology Institute Groningen (ENTEG), Faculty of Science and Engineering, University of Groningen, Groningen, The Netherlands

A. G. P. Kottapalli (✉)
Department of Advanced Production Engineering, Engineering and Technology Institute Groningen (ENTEG), Faculty of Science and Engineering, University of Groningen, Groningen, The Netherlands

MIT Sea Grant College Programme, Massachusetts Institute of Technology (MIT), Cambridge, MA, USA
e-mail: a.g.p.kottapalli@rug.nl

© The Author(s), under exclusive licence to Springer Nature Switzerland AG 2019   31
A. G. P. Kottapalli et al., *Self-Powered and Soft Polymer MEMS/NEMS Devices*,
SpringerBriefs in Applied Sciences and Technology,
https://doi.org/10.1007/978-3-030-05554-7_2

with the current scenario of different types of flexible piezoelectric energy harvesters. A comprehensive review of the recent and ongoing research work in the field of nanofiber-based energy harvesters is also presented in this chapter.

2.1 Introduction

In 1880, two scientists, Pierre and Jacques Curie, observed the piezoelectric effect for the first time. They found that when certain crystals (quartz, Rochelle salt, and tourmaline) were subjected to deformation in a particular direction, charges appeared on their opposite faces which were proportional to the amount of deformation. The main problem with naturally occurring piezoelectric materials were their low piezoelectric coefficients. In the 1950s, the discovery of lead zirconate titanate (PZT) and barium titanate (BaTiO3) exhibiting very high piezoelectric properties brought the much-needed breakthrough [1, 2]. Till now, PZT has been the most widely used piezoelectric material.

2.1.1 Mechanism of Piezoelectric Effect

Materials demonstrating piezoelectric effect have a constituent crystalline structure with no symmetry center regarding the negative or positive ions of the crystal lattice. Hence, there exists polar axis inside the crystal [3]. This can be demonstrated with the help of an α-quartz crystal in Fig. 2.1.

When the crystal is deformed along the X1-axis, a net polarization is generated along the axis. The displacement of positive and negative ions of the crystal lattice relative to each other leads to the appearance of charges on the faces perpendicular to the X1-axis which is manifested in the form of the resultant polarization (Fig. 2.2).

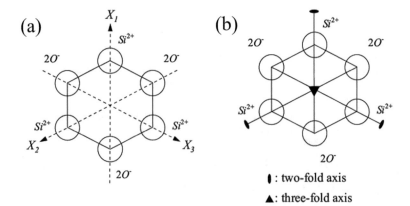

Fig. 2.1 Schematic diagram showing the structure of α-quartz. (Figure reproduced from [1] with permission of ©Springer)

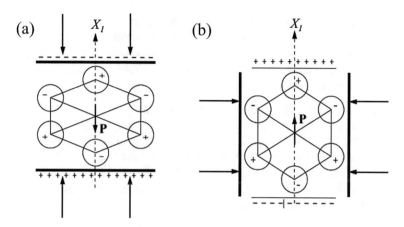

Fig. 2.2 (a) Direct longitudinal piezoelectric effect. (b) Direct transversal piezoelectric effect. (Figure reproduced from [1] with permission of ©Springer)

2.1.2   Piezoelectric Materials

As discussed earlier, the piezoelectric property is found in materials lacking inversion symmetry. Most commonly used piezoelectric materials used today can be subdivided into the following categories:

- *Lead-based piezoelectric materials*: Among all the piezoelectric materials, lead-based piezoelectric materials are most versatile and widely used. Some common lead-based piezoelectric materials are lead titanate ($PbTiO_3$ – though not used commercially, lead titanate can be modified or used to form solid solution yielding excellent piezoelectric properties) and lead zirconate titanate ($Pb(Zr_{1-x}Ti_x)O_3$ or PZT is a solid solution of $PbTiO_3$ and $PbZrO_3$ where these two compounds are soluble in all proportions) [2]. PZT is the most widely used piezoelectric material.
- *Lead-free piezoelectric materials*: Though PZT-based piezoelectric materials are most common and widely used due to their excellent piezoelectric properties, awareness of negative effects of lead on human health and overall environment has led to the exploration of alternative environment-friendly lead-free piezoelectric materials. The lead-free system can be subdivided into two categories, namely, perovskites like $BaTiO_3$ (BT), BNT, $KNbO_3$, $NaTaO_3$, etc. and non-perovskites like tungsten-bronze-type ferroelectrics, bismuth layer-structured ferroelectrics (BLSF), etc. [3].
- *Piezoelectric polymers*: This category of piezoelectric materials has gained increasing research attention recently mainly because of their mechanical property like flexibility and biocompatibility. Piezoelectric polymers can be divided into three subcategories [4]: (a) bulk piezoelectric polymers like polyvinylidene fluoride

(PVDF), polyamides, parylene-C, liquid crystal polymers, polyimide, and polyvinylidene chloride (PVDC); (b) polymer piezoelectric composites or polymer materials having inorganic piezoelectric materials embedded; and (c) voided charge polymers or polymers having internal gas voids. When the surfaces surrounding the voids get charged, these materials behave like piezoelectric materials.

In the field of MEMS/NEMS sensors, principles of piezoelectric [5–12] and piezoresistive [13–19] sensing have mostly been exploited to form sensing elements of the sensors. Piezoelectric polymers have been of great interest recently. In the past few years, piezoelectric polymers have found extensive applications in the field of biomimetic devices [20–22] and MEMS/NEMS applications [23–27]. As the core focus of this chapter is flexible self-powered sensors, the next section will deal with the piezoelectric polymers in more details.

2.2　Piezoelectric Polymers

As discussed in the previous section, the main reason behind the popularity of polymer-based piezoelectric materials is their flexibility and biocompatibility. Polymer-based piezoelectric materials have generated enough interest in the past two decades to be used as an alternative to conventional piezoceramics. Their low piezoelectric coefficients are compensated by biocompatibility and excellent mechanical flexibility. It is also important to note that a standard piezoelectric polymer like PVDF has much higher piezoelectric stress constant (g31) in comparison to piezoceramic like PZT. This implies better sensing capabilities [6]. Table 2.1 shows the comparison between PZT and PVDF representing piezoceramic and polymer-based piezoelectric materials, respectively.

Polymer-based piezoelectric materials can be divided into three broad categories: (a) bulk piezopolymers, (b) voided charged polymers, and (c) polymer piezocomposites [4]. The following schematic diagram (Fig. 2.3) shows the subcategories of piezoelectric polymers.

The individual categories are discussed briefly in the next three subsections. Though all the individual categories are discussed, the core focus of this chapter is the bulk piezopolymers as they are mostly used in flexible sensors and energy harvesters which is the main focus of this chapter.

Table 2.1 Comparing standard piezoceramic (PZT) with piezoelectric polymer (PVDF) [28]

	d_{31}, pm/V	g_{31}, (mV.m)/N	k_{31}	Salient features
Polyvinylidene fluoride (PVDF)	28	240	0.12	Excellent flexibility, biocompatibility, and lightweight
Lead zirconate titanate (PZT)	175	11	0.34	Toxic, brittle, heavy

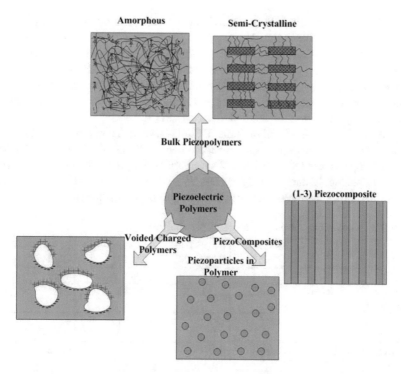

Fig. 2.3 Schematic diagram showing various categories under which the polymer-based piezo-electric materials can be categorized. (Figure reproduced from [4] with permission of ©IOPSCIENCE)

2.2.1 Bulk Piezopolymers

The piezoelectric effect is seen in bulk piezoelectric polymers due to their orientation and molecular structure. Bulk polymers can be further divided into two sub-categories: semicrystalline and amorphous polymers [28]. The following four critical requirements are to be satisfied by any polymer to exhibit piezoelectric properties [28]:

1. Permanent molecular dipoles should exist in the material.
2. It should be possible to align the molecular dipoles.
3. Once achieved, the alignment should be sustained.
4. When subjected to large stress, the material should be able to undergo large strain.

The polymers having a semicrystalline structure like polyvinylidene fluoride (PVDF) [7], parylene-C [8], liquid crystal polymers, and polyamides [6] are known for exhibiting piezoelectric properties. Their principle of operation is similar to that of conventional piezoelectric inorganic materials. The bulk of such materials contain randomly oriented microscopic crystals. By the process of poling, these crystallites are reoriented along a preferred orientation thus achieving a piezoelectric response out of the bulk [5].

Some noncrystalline polymers like polyvinylidene chloride (PVDC) [6] and polyimide [9] having molecular dipoles in their structure are also known to exhibit piezoelectric property. For these polymers, the poling is performed at a temperature which is few degrees more than the glass transition temperature which aligns the molecular dipoles along the applied electric field [5].

Among all the bulk piezopolymers, PVDF is the most commonly used polymer primarily due to its large piezoelectric coefficient of 20–28 pC N^{-1} in comparison to other piezopolymers [5, 6]. Table 2.2 summarizes the properties of the most common bulk piezopolymers, namely, PVDF, PVDF-TrFE, parylene-C, and polyimide. Figure 2.4 shows the molecular structure of these bulk piezopolymers.

Table 2.2 Summarizing the properties of bulk piezopolymers [4]

		PVDF [28, 29]	PVDF-TrFE [11, 28, 29]	Parylene-C [30, 31]	PI (β-CN) APB/ ODPA [28, 32]
Density	(kg m^{-3})	1800	1900	1290	1420
Young's modulus Y	(GPa)	2.5–3.2	1.1–3	2.8	2–3
Dielectric constant ε_r		12	12	3.15	4
Dielectric loss tan δ_e		0.018	0.018		0.001
Mechanical loss tan δ_m		0.05	0.05	0.06	0.06
d_{31}	(pC N^{-1})	6–20	6–12		
d_{33}	(pC N^{-1})	13–28	24–38	2.0	5.3–16.5
k_{31}		0.12	0.07		
k_{33}		0.27	0.37	0.02	0.048–0.15
Maximum use temp.	°C	90	100		220

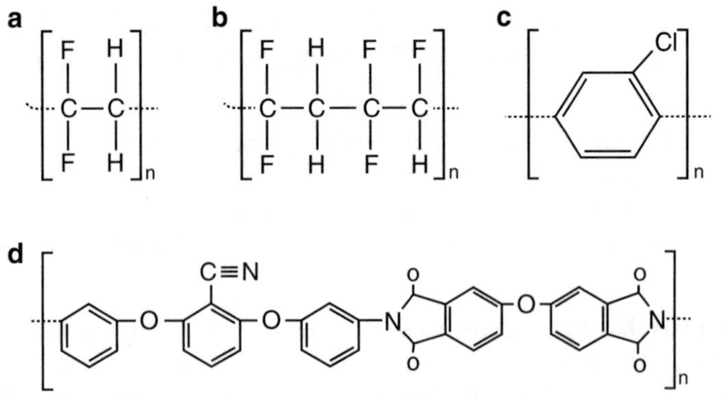

Fig. 2.4 Diagram showing the molecular structure of bulk piezoelectric polymers: (**a**) PVDF, (**b**) PVDF-TrFE, (**c**) parylene-C, and (**d**) PI (β-CN) APB/ODPA. (Figure reproduced from [4] with permission of ©IOPSCIENCE)

2.2.2   Polymer Piezoelectric Composites

A piezo-composite is a material in which inorganic piezoelectric materials are embedded inside a polymer material. In this type of materials, the polymer is non-electroactive [4]. Mixing polymers with piezoelectric polymers helps in combining the excellent mechanical flexibility offered by the polymer with the large piezoelectric coefficient and dielectric constant of the ceramic.

Newnham et al. discussed microstructural arrangement of component phases in composites (also known as connectivity) which was later amended by Pilgrim et al. [33, 34]. There are 16 possible connectivity patterns for composite consisting of two phases. They vary from [0-0] (which means neither of the phases is self-connected) to [3] (each phase is self-connected in three dimensions) [35]. Companies like Smart Materials commercially manufacture [1–3] composites where the ceramic rods are arranged or scattered randomly in polymer bulk films [36]. For MEMS applications, the most commonly used configuration is rods or particles embedded in bulk polymer films [4].

2.2.3   Voided Charged Polymers

Voided polymer electret was first used in 1962 when G.M. Sessler and J.E. West from Bell Telephone Laboratories invented a condenser microphone with a solid dielectric between its two electrodes [37]. It was considered as "space charged electrets." It was not until the 1980s that researchers started investigating the pyro- and piezoelectricity of such films [38].

Voided charged polymers have gas voids embedded inside of the polymer film. These materials behave like piezoelectric materials when the polymer surfaces surrounding the gas voids get charged. The poling process of voided charge polymers is similar to that of the conventional electrical poling process followed in case of bulk piezoelectric polymers (Fig. 2.5). The process starts with a polymer film with embedded air voids which is subjected to electrical poling. Upon the application of a certain electric field, the gas molecules inside the voids get ionized. The ionized molecules accelerate in the direction of the applied electric field and get implanted in the wall of the voids [38]. These voids act like artificial dipoles which respond to an externally applied electric field or mechanical force imitating a conventional piezoelectric material.

Some common voided charged polymers found in the literature [4]:

- *Void formation and expansion-based VCPs*: Cellular polypropylene, fluorinated and post-treated cellular polypropylene, COC-based cellular electrets, and cellular polyethylene-naphthalate (PEN)
- *Multilayer VCPs*: PTFE/FEP multilayer VCP, FEP multilayer, cellular PDMS, and micromachined integrated cellular parylene

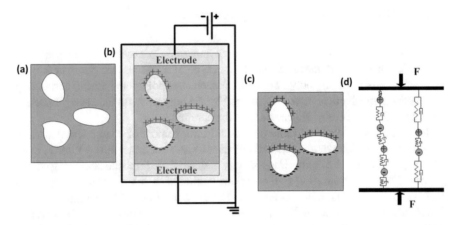

Fig. 2.5 Schematic diagram showing piezoelectricity in voided charge polymers: (**a**) polymer with embedded air voids before charging, (**b**) poling process to form the dipoles, (**c**) equivalent model to explain the piezoelectric response of the voided charged polymer. (Figure reproduced from [4] with permission of ©IOPSCIENCE)

2.3   Piezoelectric Energy Harvesters

Due to a surge in the demand for wearable and portable devices with long-lasting battery lives, the field of energy harvesting has seen a massive growth over the past two decades. Though there has been a lot of progress in the field of conventional power sources used in portable devices in the form of introduction of lithium ion and lithium polymer batteries with high energy packing density, there still remains a huge scope for improvement in the present scenario. Both academia and industry are now trying to develop efficient energy harvesters to supplement the conventional power sources in order to increase the battery life of wearable and portable devices. Also, there are environments and applications where periodic charging and replacement of batteries might not be possible. The pacemaker is one such device where changing the battery is difficult and poses a serious risk to the host. With the advancement of technology, wireless implants are also becoming popular where the power sources have to be long lasting which raises a serious question on their practicability.

Considering all these scenarios, it is safe to assume that energy harvesters will be the future of portable and wearable devices. Simply put, energy harvesters are devices which convert the ambient energy surrounding it into electrical energy which can be used for powering various devices. There are different types of macro-scale energy harvesters known to us like thermal power, hydroelectric power, solar power, biogas, etc. In the context of microsystems and biomedical applications, piezoelectric energy harvesting is the most preferred method for energy scavenging [39].

The main focus of this section is to discuss various types of piezoelectric energy harvesters developed so far. For this purpose, we have segregated the energy harvesters into two subcategories: piezoelectric thin film/bulk energy harvesters and piezoelectric nanofiber energy harvesters. In the following subsections, each of them is discussed separately.

2.3.1 Piezoelectric Thin Film/Bulk Energy Harvesters

Piezoelectric thin film/bulk energy harvesters for wearable/implantable systems have been widely researched in the past two decades. Gonzalez et al. carried out an extensive study on the energy harvesting potential from the human body [40]. They divided human activities into two subcategories, namely, continuous and discontinuous activities. Activities like blood flow and chest expansion during breathing were grouped under the continuous category, whereas upper limb movements and walking were grouped under the discontinuous category. They calculated that continuous activities like blood flow generate 0.93 watts, and chest expansion can generate up to 0.83 watts of power. However, it is important to note that harnessing all of the 0.93 watt power generated during blood flow will hinder the normal functioning of the heart. In contrast, discontinuous activities like finger movement during typing (generates 19 mW), upper limb motion during regular activities (generates 3 W), and walking (generates 67 W) generate significantly more power. Another study carried out by Niu et al. found that up to 2.2 W, 2.1 W, 39.2 W, 49.5 W, and 69.8 W of power could be generated from the shoulder, elbow, hip, knee, and ankle motion, respectively [41]. After all the evaluations, the authors concluded that placing of piezoelectric energy scavenger inside a shoe sole qualified as the best candidate for harvesting energy from human body motion.

The fundamental characteristic of human motion is low-frequency large amplitude movements which pose some serious challenges in the design of miniature resonant generators [42]. Hence, for human-centric applications, mechanical to electrical transduction is usually achieved by direct straining of the piezoelectric element.

Previously, research has been carried out at MIT media laboratory to study the amount of energy dissipated while walking. It was found that for an average human being weighing 68 Kg with average gait, 67 watts of power was generated at the heel [43]. Here, it is important to note that extracting this amount of power will interfere with a person's gait. Nonetheless, the aforementioned study showed the potential of parasitic energy harvesting in shoes.

Two piezoelectric energy harvesters were developed at MIT media laboratory which were used for demonstrating the feasibility of harnessing useful power from the shoes (Fig. 2.6) of a moving person [44]. The first system harnessed the energy generated during the bending of the sole. The system consists of a "stave" which is bimorph structure where the 2 mm-thick central flexible plastic core is sandwiched between eight-layer stacks of electrode-laminated PVDF with thickness of 28 μm (Fig. 2.7). This "stave" was placed inside the sole of a sports sneaker where the bending of the sole caused strain in the PVDF stacks hence producing charge (d_{31} mode). This setup delivered an average power of 1.3 mW to a 250 kΩ load at a footfall rate of 0.9 Hz.

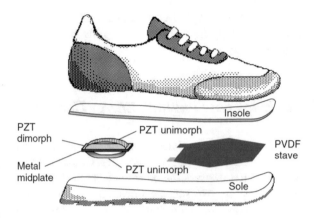

Fig. 2.6 Diagram showing the placement of the two discussed insole energy scavengers. (Figure reproduced from [45] with permission ©IEEEXplore)

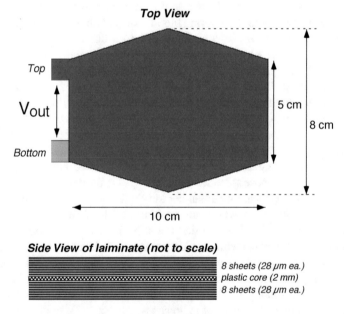

Fig. 2.7 Schematic diagram showing the design of PVDF "stave". (Figure reproduced from [44] with permission ©IEEEXplore)

The second system which is referred to as dimorph consists of spring steel bonded to a flexible piezoceramic patch. This method harvests the energy dissipated under the heel by flattening curved, prestressed spring metal strips laminated with a semiflexible form of PZT [45]. This dimorph structure consisted of two piezoelectric transducers (Thunder TH-6R) made by Face International Corporation [46]. The structure is designed such that there is a difference of thermal expansion coefficients in the materials which leads to a curved structure (Fig. 2.8). This curved structure leads to a com-

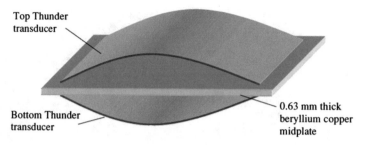

Top Thunder
transducer

Bottom Thunder
transducer

0.63 mm thick
beryllium copper
midplate

Fig. 2.8 Schematic diagram showing the design of PZT dimorph. (Figure reproduced from [42] with permission ©IOPSCIENCE)

pressive stress in the PZT layer which allows it to bend much more than conventional PZT structures. The transducer is deformed with each taken step when the heel hits the ground and returns to its original shape when the heel is lifted. This setup delivered an average power of 8.4 mW to a 500 kΩ load at a footfall rate of 0.9 Hz. Due to these research efforts, today insole energy harvesting is becoming mainstream.

Sohn et al. carried out a detailed study where they conducted theoretical modeling and finite element method (FEM) simulations to evaluate the practicality of using commercially available piezoelectric films for harnessing energy from fluctuating pressure source such as human blood pressure [47]. They modeled and compared several circular and square PVDF films. It was concluded that the maximum power (0.61 μW) was generated by a circular PVDF film of radius 5.62 mm with a thickness of 9 μm when subjected to a uniformly distributed pressure (5333 N m^{-2}) equivalent to human blood pressure. The result was experimentally validated using a setup consisting of a 28 μm-thick circular PVDF film placed inside an aluminum jig having inlet and outlet ports for pumping in and pumping out liquid (Fig. 2.9a, b). A pneumatic compressor was used to supply the uniform pressure of 5333 N m^{-2} emulating human blood pressure. A digital oscilloscope was used for measuring the voltage generated by the PVDF film. When subjected to a sinusoidal (1 Hz) pressure of 5333 N m^{-2}, the film produced 0.33 μW of power.

Platt et al. demonstrated the application of piezoelectric ceramics in energy harvesting from total knee replacement (TKR) units [48]. An extensive study was carried out to model the energy harvester in total knee replacement unit where factors such as efficiency, longevity, energy conversion, load matching, form factors, and energy storage were taken into consideration. The model was experimentally validated with a prototype total knee replacement unit (Fig. 2.10) using a force profile of 2600 N which emulates the axial force generated in human knees while walking. A raw power of 4 mW was generated. The maximum regulated power generated from the setup was 0.85 mW.

Cantilever geometries with piezoelectric materials attached on the top and bottom are most suitable for vibrational energy harvesting due to their low resonant frequencies which can be further reduced by attaching proof masses at the tips [42]. They are designed to operate in the d$_{31}$ mode where the piezoelectric materials are strained when the cantilevers are bent.

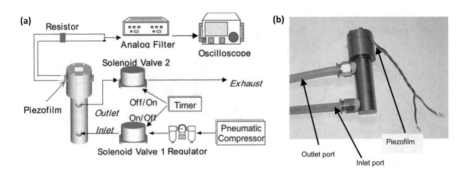

Fig. 2.9 (**a**) Schematic diagram showing the experimental setup, (**b**) diagram showing the jig. (Figure reproduced from [47] with permission ©Sage journals)

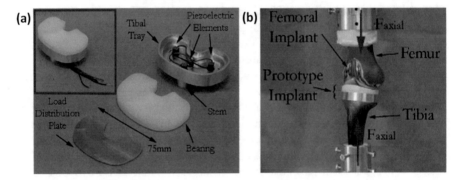

Fig. 2.10 (**a**) Individual components of the self-powered TKR unit, (**b**) test setup for the total knee replacement unit. (Figure reproduced from [48] with permission ©IEEEXplore)

Roundy and Wright developed and evaluated a vibration-based piezoelectric energy harvester for wireless network applications [49]. Their basic design consists of a bimorph structure with a proof mass attached at the free end (Fig. 2.11).

Their argument for choosing a bimorph cantilever structure was mainly due to two reasons:

1. The cantilever configuration leads to the highest value of average strain for a particular input force. The average strain directly impacts the power output.
2. For a particular size, the cantilever configuration results in the lowest resonant frequency.

A detailed mathematical analysis was carried out by the authors [49]. First, a prototype generator was fabricated which consisted of a central steel shim sandwiched between two PZT-5A layers. An alloy mass (made of tin and bismuth) was attached to the end of the bimorph cantilever. The prototype was tested by subjecting it to a 2.5 ms^{-2} vibration at a frequency of 120 Hz. The power generated was

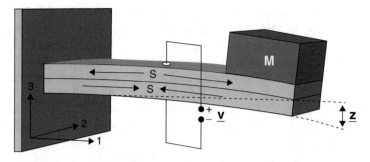

Fig. 2.11 Schematic diagram showing the design of the bimorph cantilever energy harvester. (Figure reproduced from [49] with permission ©IOPSCIENCE)

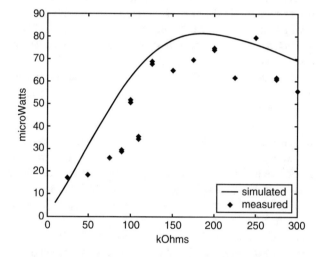

Fig. 2.12 Plot comparing the simulated values with experimental data for power delivered as a function of load resistance. (Figure redrawn from [49])

plotted as a function of load resistance (Fig. 2.12). The agreement between the simulated results and measured data was deemed sufficient by the authors for using the basic design as a stepping stone for further optimizations. Two designs having an overall volume constraint of 1 cm^{-3} were developed using PZT-5H (with brass as the center shim). For a detailed description of the optimized designs, the readers are referred to [49].

Jeon et al. developed a MEMS-based energy harvester and named it as "piezoelectric micro power generator (PMPG)" [50]. The harvester developed by the team was based on thin film PZT as the piezoelectric layer operating in d_{33} mode. The device consisted of a flat cantilever structure with a proof mass attached to its tip. In order to make the device operate in the desired d_{33} mode, the top Pt/Ti electrode was patterned into an interdigitated layout on top of the spin-coated sol-gel PZT thin film. For detailed fabrication process, the readers can refer to the work by the authors presented in [50]. The PMPG was excited at its resonant frequency (13.9 kHz) which generated a dc voltage of 3 V across a resistive load (10.1 MΩ). Vibrating at the same resonant frequency, the energy harvester delivered a maximum of 1 µW continuous power to a 5.2 MΩ load at a dc voltage of 2.4 V developed across the load. The energy density of the device was reported as 0.74 mW h/cm^2, which is comparable to that of Li-ion batteries. The voltage generated by the device (operating in d_{33} mode) was found to be 20 times higher than the voltage generated by an equivalent cantilever design operating in d_{31} mode [50].

Lee et al. developed a novel fabrication technique involving aerosol deposition method to fabricate two different PZT thin film-based MEMS energy harvesters operating in d_{31} and d_{33} modes [51]. For the details regarding the aerosol deposition technique, the readers could refer to the authors' works in articles [51–53]. The energy harvester operating in d_{31} mode generated a maximum open-circuit voltage of 2.675 V_{P-P} with a maximum power output of 2.765 µW at 1.792 V_{P-P} when excited at its resonant frequency (255.9 Hz). Whereas, the energy harvester operating in d_{33} mode generated a maximum open-circuit voltage of 4.127 V_{P-P} with a maximum power output of 1.288 µW at 2.292 V_{P-P} when excited at its resonant frequency (214 Hz). A comparison was also made by exciting both the devices at an acceleration value of 2 g (see Table 2.3).

Aktakka et al. presented the fabrication and testing of a bulk piezoelectric ceramic-based CMOS-compatible energy harvester [54]. The main advantages of this bulk PZT-based technology over other piezoelectric thin film technologies are fabrication flexibility and enhanced device performance due to bulk piezoelectric properties. They were able to obtain 5 µm to 100 µm-thick films without any chemical patterning. The fabricated energy harvester generated a power output of 0.15 µW when excited with 0.1 g acceleration at 263 Hz. The same harvester generated 10.2 µW power when excited with 2 g acceleration at 252 Hz.

Though electroactive polymers like PVDF are very flexible and easy to process, their low piezoelectric d_{31} and d_{33} coefficients limit their usability in energy harvesting applications. To overcome this problem, inorganic fillers like barium titanate, PZT, lead titanate, etc. can be added to PVDF matrices. In their work, Rahman et al. reported the energy harvesting capabilities of PVDF graphene nanocomposite-based films [55]. Nanocomposites of PVDF and graphene oxide (PVDF-GO) and PVDF and reduced graphene oxide (PVDF-RGO) made by in situ thermal reductions of graphene oxide were analyzed in terms of ferroelectric properties and energy harvesting capabilities. The three different energy harvesters were made by attaching the PVDF, PVDF-GO, and PVDF-RGO to a 400 µm-thick cantilever beam of length

Table 2.3 Comparison of the two devices excited at 2 g acceleration [51]

Operation mode	Resonant frequency	Optimal load	Power output	Output voltage (open circuit)	Output voltage (with load)
d_{31}	255.9 Hz	150 kΩ	2.099 µW	2.415 V	1.587 V
d_{33}	214.0 Hz	510 kΩ	1.288 µW	4.127 V	2.292 V

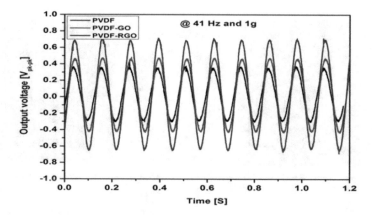

Fig. 2.13 Plot comparing the peak-to-peak voltages generated by the PVDF, PVDF-GO, and PVDF-RGO films. (Figure reproduced from [55] with permission ©IOPSCIENCE)

25 mm and width 10 mm made of FR-4 material. The peak-to-peak voltages generated by the three devices were compared (Fig. 2.13) by exciting the devices with an acceleration level of 1 g at 41 Hz. The PVDF-RGO-based cantilever showed the best peak-to-peak performance of 1.3 volts. The PVDF-RGO film-based energy harvester generated a power of 36 nW when connected to a 704 kΩ resistive load.

Won et al. developed a piezoelectric poly(vinylidene fluoride trifluoroethylene) (P(VDF-TrFE))-based flexible power generator fabricated using cellulose paper as the substrate [56]. The final structure of the energy harvester consisted of Pt (150 nm)/P(VDF-TrFE) (1 μm)/Pt (200 nm)/cellulose paper (200 μm). To demonstrate the practical application of the energy harvester, the flexible energy harvester was placed on the back of a hand and was secured using a latex glove (Fig. 2.14a). The output of the harvester was measured while gripping and releasing the hand periodically at two different frequencies (0.25 Hz and 2 Hz). The harvester generated 0.4 V and 0.6 V at 0.25 Hz and 2 Hz, respectively (Fig. 2.14b, c). The authors also reported having generated a maximum open-circuit voltage of 1.5 V and short-circuit current of 0.38 μW implying a power density of 2.85 mW/cm³ while bending the device by 0.7 cm at a frequency of 1 Hz.

2.3.2 Piezoelectric Nanofiber Energy Harvesters

Other than the thin film/bulk-based piezoelectric nanogenerators/energy harvesters, there is another class of energy harvesters made out of electrospun piezoelectric nanofibers. Electrospun nanofibers have been of important research interest for the last three decades. A significant amount of progress has been made in fabrication technologies and applications related to electrospinning. The main features of

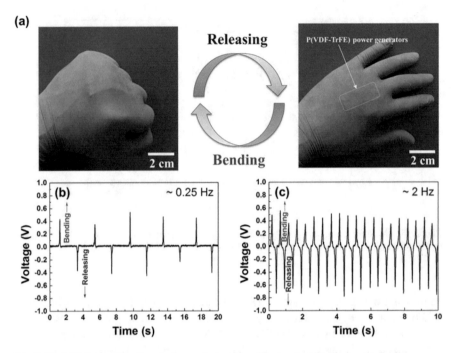

Fig. 2.14 (**a**) Images of cyclic gripping and releasing of the hand while placing the flexible energy harvester on the back of the hand under the glove. (**b**) Voltage generated at 0.25 Hz periodic hand movement. (**c**) Voltage generated at 2 Hz periodic hand movement. (Reprinted from [56], with the permission of AIP Publishing)

nanofibers which make them attractive for a wide range of applications are their large surface area to volume ratio, superior mechanical properties like tensile strength and stiffness, and excellent flexibility [57]. Due to their superior mechanical performance and ease of fabrication, many research groups all over the world have engaged in electrospinning piezoelectric materials for fabricating piezoelectric nanofibers for self-powered sensors and energy harvesters for the last two decades. The two most attractive features of electrospinning which makes it popular for fabricating piezoelectric nanofibers are mechanical stretching and in situ poling associated with the electrospinning process. Due to the intense stretching of the released polymer jet during electrospinning process in the presence of a high electric field, the nanofibers are poled in situ thus circumventing the electrode poling process as seen in case of piezoelectric thin films/bulks. The process of electrospinning is well researched and has been discussed by a lot of review articles in the past [57–61]. Of all the piezoelectric nanofibers, polyvinylidene fluoride-based nanofibers have been widely used for energy scavenging applications [62]. Nanofibers can be fabricated by either of the two most popular electrospinning processes:

- *Near-field electrospinning (NFES)*: This method is suitable for fabricating and depositing a single continuous nanofiber in a controlled way [63].

- *Far-field electrospinning (FFES)*: This is the more conventional method which is suitable for producing a large mat of dense nanofibers [57].

The main difference between the NFES and FFES lies in the needle tip to collector distance. While NFES allows the needle tip collector gap as low as 1 mm, the corresponding gap for the FFES is usually more than 10 mm [5]. This section discusses some of the piezoelectric nanofiber-based energy harvesters fabricated by the process of electrospinning.

PZT-based energy harvesters exhibit better power output and maximum voltage in comparison to other piezoelectric materials for a given volume of the material. However, like all other ceramic materials, the main problem with PZT thin film/bulk-based structures is their extreme fragility. Thin films which are very sensitive to vibrations of high frequency face even higher risk of failure due to breakage [64]. Earlier, Chen et al. have successfully demonstrated the capability of PZT nanofibers to solve the problems associated with the thin films/microfibers by exhibiting high mechanical flexibility and mechanical strength while demonstrating a high value of piezoelectric voltage constant (g_{33}, 0.079 Vm/N) [65]. In another work, the same research group demonstrated a PZT nanofiber-based energy harvester [64]. Laterally aligned PZT nanofibers were placed on interdigitated electrodes made of fine platinum wire arrays of 50 μm diameter assembled on a silicon substrate (Fig. 2.15). The diameters of the nanofibers were reported to be 60 nm, and the gap between two adjacent electrodes was kept at 500 μm. The nanofibers were poled separately by applying an electric field of 4 V/μm across the electrodes at an elevated temperature of 140 °C for 24 h. A maximum output voltage of 1.63 V was achieved during the tests. The nanogenerator was able to generate a maximum power of 0.03 μW when connected to a 6 MΩ resistive load.

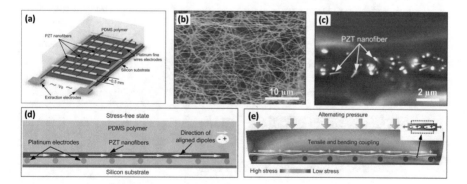

Fig. 2.15 (**a**) Schematic diagram of the layout of the energy harvester showing the PZT nanofibers placed on interdigitated electrodes. (**b**) SEM images of electrospun PZT nanofiber mat. (**c**) SEM image showing the cross-sectional view of the PZT nanofibers embedded in PDMS matrix. (**d**) Schematic diagram showing cross-sectional view of the final energy harvester structure with PDMS encapsulation. (**e**) Schematic diagram explaining the charge generation mechanism of the energy harvester. (Reprinted (adapted) with permission from [64]. Copyright (2010) American Chemical Society)

The two main problems associated with the use of PZT nanofibers for energy scavenging applications are:

- The requirement of high temperature (~ 650 °C) annealing to achieve pure perovskite phase [64].
- For electrospinning, PZT has to be mixed with a suitable solvent which lowers the PZT content in the electrospun nanofibers which in turn leads to the lowering of piezoelectric coefficient thus lowering the overall energy scavenging capability of such nanofibers [62].

In comparison, polyvinylidene fluoride-based piezoelectric nanofibers offer a unique combination of mechanical flexibility, biocompatibility, and lightweight coupled with an easy fabrication process (which does not require high-temperature annealing or poling) making them an ideal candidate for fabricating piezoelectric nanofiber-based energy harvesters for wearable devices and human implant applications.

Chang et al. made use of the direct-write technique by means of near-field electrospinning to place PVDF nanofibers on substrates (Fig. 2.16) in a controlled fashion [66]. As discussed earlier, the nanofibers are mechanically stretched and poled

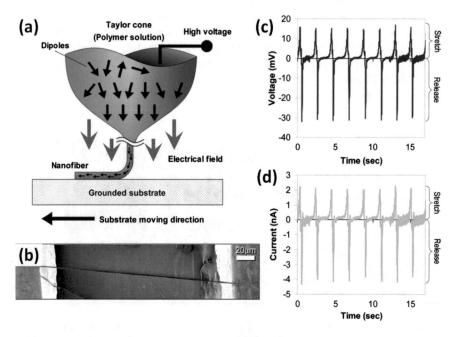

Fig. 2.16 (**a**) Schematic diagram illustrating the NFES to place PVDF nanofiber on a substrate. (**b**) SEM image showing the top view of the nanogenerator consisting of a single PVDF nanofiber placed on two separate electrodes deposited on a plastic substrate. (**c**) Voltage generated by the nanogenerator under a cyclic (frequency of 2 Hz) strain. (**d**) Current generated by the nanogenerator under a cyclic (frequency of 2 Hz) strain. (Reprinted (adapted) with permission from [66]. Copyright (2010) American Chemical Society)

in situ during the electrospinning process which leads to the alignment of the dipoles in the nanofiber crystals thus leading to transformation of nonpolar α-phase to the polar β-phase (interested readers may refer to some of the works where detailed studies have been carried out to find out the effect of various parameters on phase transformation in PVDF thin films and nanofibers [5, 67]). Nanofiber diameters of 500 nm to 6.5 μm were reported. The length of the nanofibers depended on the separation (100–600 μm) between the metallic electrodes. A voltage output in the range 5–30 mV and a current output of 0.5–3 nA was reported after testing of 50 nanogenerators. A highest conversion efficiency of 21.8% and an average conversion efficiency of 12.5% were observed for 45 tested nanogenerator samples. The authors also reported having enhanced the output voltage and current of the nanogenerators by connecting them serially and parallelly, respectively.

Hansen et al., for the first time, demonstrated the integration of PVDF-based nanogenerator with biofuel cell for in vivo energy harvesting applications [23]. A modified far-field electrospinning process was adopted for electrospinning the PVDF nanofibers on a kapton film (Fig. 2.17a). The split electrode approach helps in aligning the nanofibers [68]. The nanofibers were secured with silver paste. As the fabrication method was far-field electrospinning process, the nanofibers had to be poled separately with an electric field of 2 MV/cm for 15 min using an in-plane electrode poling process. The nanogenerator was encapsulated with PDMS (Fig. 2.17b). The PVDF nanogenerator was integrated with a biofuel cell (Fig. 2.17c). The nanogenerator generated a maximum voltage of 20 mV and a maximum current output of 0.3 nA with a fixed strain rate of 1.67%/s (Fig. 2.18).

As seen in the previous section on piezoelectric thin film/bulk energy harvesters, copolymers of PVDF have also been used to fabricate nanofiber-based energy harvesters. Mandal et al. reported electrospun poly(vinylidene fluoride-trifluoroethylene) (P(VDF-TrFE)) nanofiber-based pressure-type nanogenerator [69]. A far-field electrospinning process with rotating mandrel collector was used for the electrospinning of the PVDF-TrFE nanofibers. A maximum voltage of 400 mV was reported to have been generated by the fabricated nanogenerator consisting of 43 μm-thick nanofiber web under periodic pressure tests with 0.2 MPa pressure at 5.3 Hz frequency. Nanofiber webs were also serially connected to either double the total voltage output or cancel out thus reducing the total voltage output of the nanofiber webs depending on their polarity (Fig. 2.19b).

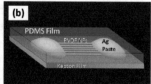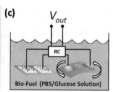

Fig. 2.17 Schematic diagram of (**a**) the modified FFES using split electrode method. (**b**) The nanogenerator with PDMS encapsulation. (**c**) PVDF nanogenerator and biofuel cell integration. (Reprinted (adapted) with permission from [23]. Copyright (2010) American Chemical Society)

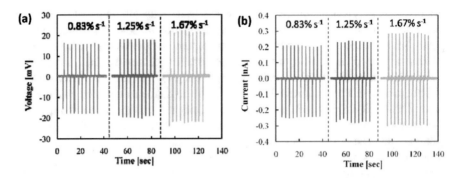

Fig. 2.18 Plot showing: (**a**) The open-circuit voltage. (**b**) The short-circuit current of the nanogenerator under different strain conditions. (Reprinted (adapted) with permission from [23]. Copyright (2010) American Chemical Society)

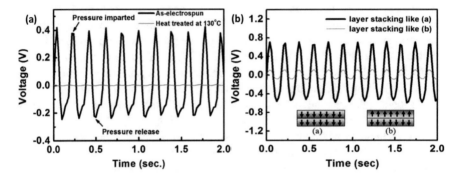

Fig. 2.19 Plot showing: (**a**) The output of the nanofiber web-based generator before and after annealing at 130 °C. (**b**) The output of serially connected nanogenerators where the output depends on polarity of individual nanofiber webs. (Figure reproduced from [69] with permission ©Wiley-VCH)

Liu et al. reported a hollow cylindrical near-field electrospinning (HCNFES) process to fabricated nonwoven well-aligned PVDF nanofibers having diameters in the range 200 nm to 1.16 μm (for different applied electrospinning voltages) [70]. The so-called hollow cylindrical near-field electrospinning (HCNFES) process consisted of a rotating glass tube collector with copper foil placed inside the internal wall grounded with an electrical brush. Figure 2.20 shows the details of the electrospinning setup. The nonwoven nanofiber fabric was transferred to a polyethylene terephthalate (PET) substrate with copper foil electrodes and secured using silver paste to make a PVDF nanofiber-based energy harvester (Fig. 2.21a). The assembly of substrate and the nanofiber fabric was covered with a thin polymer layer. The assembly was subjected to a 7 Hz periodic stretching and releasing with a strain of 0.05% using the setup shown in Fig. 2.21b. The nanogenerator produced a maximum power of 577.6 pW cm^{-2}, an average peak voltage output of −76 mV (across a 10 MΩ resistive load), and −39 nA peak current.

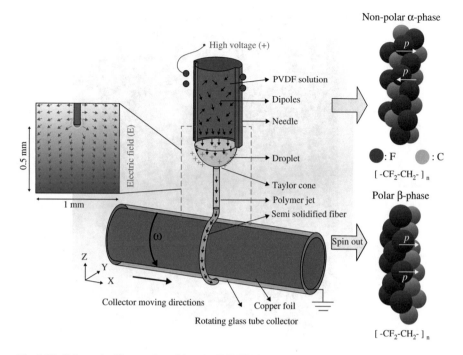

Fig. 2.20 Schematic diagram describing the HCNFES process. Due to in situ electric poling and intense stretching of polymer jet, the phase transformation (nonpolar α-phase to polar β-phase) of PVDF takes place. (Figure redrawn from [70])

2.3.3 Piezoelectric Nanowire Energy Harvesters

In the past few years, a significant amount of research has been carried out to develop energy harvesters employing piezoelectric nanowires. The features of nanowires which make them attractive for energy scavenging applications are their exceptional sensitivity to small random mechanical disturbance/stimulation, high mechanical robustness, enhanced flexibility, and lightweight [71, 72]. In this section, a brief overview of piezoelectric nanowire-based energy harvesters developed by various research groups across the world is presented.

Among all the one-dimensional nanomaterials, zinc oxide is preferred because of the following reasons [73]:

- ZnO is most suitable for electromechanically coupled sensor applications because of its unique combination of piezoelectric and semiconductor properties.
- Zinc oxide is relatively bio-safe and can be used for biomedical applications (with little toxicity).
- ZnO is known to exhibit most diverse configurations.

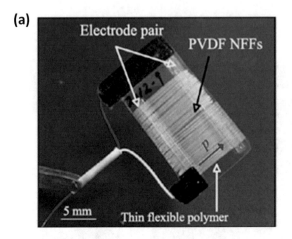

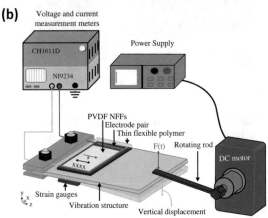

Fig. 2.21 (**a**) The fabricated nanogenerator. (Figure reproduced from [70] with permission ©IOPScience) (**b**) Experimental setup for testing the nanogenerator. (Figure redrawn from [70])

Wang and Song carried out a pioneering work in the field of nanoscale energy harvesters when they converted mechanical energy to electrical energy using an array of piezoelectric zinc oxide nanowires [73]. An array of relatively short (0.2 μm to 0.5 μm) aligned ZnO nanowires were grown on c plane-oriented α-Al$_2$O$_3$ by vapor-liquid-solid (VLS) process (Fig. 2.22a, b). The outputs of the nanowires were measured using an atomic force microscope (AFM) with a platinum-coated silicon tip having a 70 ° cone angle (Fig. 2.22c). The output voltage generated by the nanowires across a 500 MΩ resistive load was observed when the AFM tip deflected the nanowires while scanning. The measurements were reported to have been carried out in AFM contact mode with a normal force of 5 nN maintained between the AFM tip and sample surface. Though a practical application was not demonstrated in this work, the authors theoretically estimated the feasibility of such nanowire

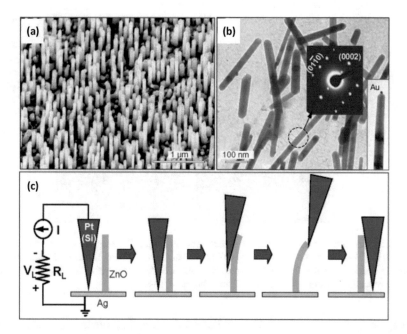

Fig. 2.22 (a) SEM images showing aligned zinc oxide nanowire array on α-Al$_2$O$_3$ substrate. (b) TEM images showing structure of individual zinc oxide nanowire with/without gold nanoparticle on top. Inset: showing electron diffraction pattern from a ZnO nanowire. (c) AFM experimental setup for testing the voltage output from individual nanowires. (Figure reproduced from [73] with permission ©SCIENCE)

nanogenerator for powering nanodevices. The power density of the nanowire array was estimated as 10 pW/μm^2 assuming that the density of nanowires per unit surface area is 20/μm^2 and a nanowire typically has a resonant frequency of 10 MHz thus generating a power output of 0.5 pW each. It was thus concluded that the power generated by a 10 μm × 10 μm nanowire array would be sufficient for powering a nanotube/nanowire-based device.

Wang et al. expanded on their previous work [73] to drive all the individual zinc oxide nanowires simultaneously using ultrasonic wave by means of a zigzag metal electrode in order to have a practical nanowire-based energy harvester capable of generating a continuous current [74]. The zinc oxide nanowires grown on GaN substrate were covered with a zigzag silicon electrode coated with platinum which increased the conductivity of the electrode and created a Schottky contact at the interface (Fig. 2.23). The density of the nanowires per unit surface area was reported to be 10/μm^2. The height and diameter of the individual nanowires were reported as ~1 μm and ~40 nm, respectively. The packaged device was driven by an ultrasonic wave of 41 kHz frequency, and the output current and voltage were measured. The voltage and current output of the nanogenerator were monitored by turning on and off the ultrasonic wave. When the ultrasonic wave was turned on, the output exhibited a ~0.15 nA jump from baseline current. A similar result was observed for the

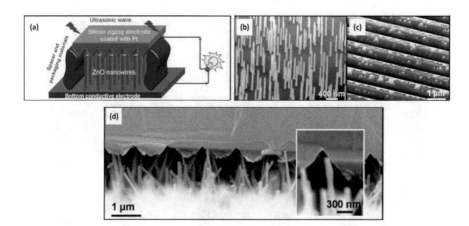

Fig. 2.23 (**a**) Schematic diagram of the nanogenerator with Si electrode coated with Pt. (**b**) Aligned ZnO nanowires on gallium nitride (GaN) substrate. (**c**) Zigzag Si electrode fabricated by standard etching process. (**d**) SEM image showing cross-sectional view of the nanogenerator. (Figure reproduced from [74] with permission ©SCIENCE)+

voltage output with ~ −0.7 mV drop. For details regarding the mechanisms behind the negative voltage drop, the readers are referred to the original work [74].

Driving a nanogenerator with AFM tips or ultrasonic waves has limitations in practical applications involving energy scavenging from human bodily movements like heartbeat, footsteps, etc. The same research group which previously developed the ultrasonic wave driven d.c. nanogenerator came up with a novel nanogenerator made up of piezoelectric zinc oxide nanowires grown radially around Kevlar 129 fibers [75].

The basic nanogenerator consists of two entangled Kevlar 129 fibers with single crystalline zinc oxide nanowires (with a typical length ~3.5 μm and diameter ~ 50–200 nm) grown radially all around using hydrothermal approach (Fig. 2.24). The gap between the individual nanofibers grown on the textile fiber surface is of the order of few hundred nanometers. The continuous zinc oxide layer on the surface of the Kevlar fiber acted as the common electrode connecting the bottom of all ZnO nanowires for electrical contact. One of the two fibers is coated with 300 nm layer of gold, and the other is left in the as-grown state. Figure 2.25a shows the schematic diagram of the double fiber nanogenerator system. The gold-coated fiber was attached to an external rotor by means of a string to stimulate relative brushing motion between the two Kevlar fibers. Here, it is worth observing that the gold-coated zinc oxide nanowires on one of the two Kevlar fibers emulate an array of scanning metal tips which in principle is similar to the zigzag silicon electrode seen in the previous case of ultrasonic wave-driven energy harvester. Figure 2.25c, d shows the schematic illustrations explaining the charge generation principle of the nanogenerator. The total output current from the system is the sum of the individual currents generated by the individual zinc oxide nanowires. The rotor was made to rotate at 80 r.p.m. to stimulate stretching and releasing of the gold-coated fiber (thus

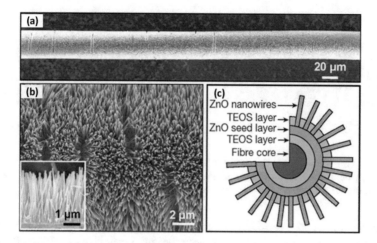

Fig. 2.24 (**a**) Scanning electron microscopy image showing a zinc nanowire-coated Kevlar 129 fiber. (**b**) Magnified scanning electron microscopy image showing zinc nanowires on Kevlar fiber. (**c**) Illustration showing cross-section of zinc nanowire-coated TEOS-enhanced Kevlar fiber. (Figure reproduced from [75] with permission ©Nature Publishing Group)

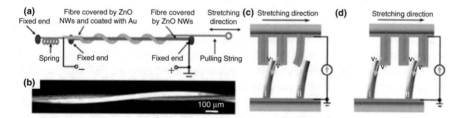

Fig. 2.25 (**a**) Schematic diagram explaining the working of a basic two-fiber nanogenerator. (**b**) Optical image of two entangled Kevlar fiber (one of which is coated with gold). (**c**), (**d**) Schematic diagram showing the teeth-to-teeth contact between the gold-coated nanowires and the as-grown nanowires leading to charge generation. (Figure reproduced from [75] with permission ©Nature Publishing Group)

leading to a constant scrubbing between two fibers), and open-circuit voltage and short-circuit current were continuously monitored. Here, the authors reported to have used a "switching polarity" method for the measurements where the current and voltmeter are both forward and reverse connected (w.r.t. polarity) to the nano-generator to weed out any system artifact.

For forward-connected current meter, a ~5 pA current signal was observed for each pull release cycle (blue line in Fig. 2.26a). For reverse connection, a current output of ~ −5 pA was observed (pink line in Fig. 3.26a). The same "switching polarity" method was used for measuring the open-circuit voltage, and a voltage amplitude of ~1–3 mV was observed (Fig. 2.26b). The authors also reported having investigated approaches for increasing the overall power generation capacity of the

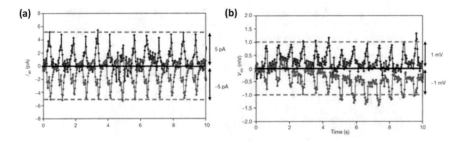

Fig. 2.26 Plots showing: (**a**) Output short-circuit current of the two Kevlar fiber nanogenerator setup. (**b**) Output open-circuit voltage of the two Kevlar fiber nanogenerator setup. (Figure reproduced from [75] with permission ©Nature Publishing Group)

nanogenerator. A test consisting of three pairs of nanowire-coated Kevlar fibers emulating a real yarn was carried out with the same conditions as that of the basic experiment consisting of a single pair of Kevlar fibers. The current output was increased to ~0.2 nA.

2.4　Conclusions and Future Work

The last two decades have seen an immense rise in the popularity of portable and wearable smart devices. With the growing popularity of portable devices, the need for long-lasting power sources has become an absolute necessity. Reliable batteries coupled with efficient energy harvesters will pave the way for long-lasting portable devices and self-powered sensors. In this chapter, a systematic review has been carried out covering some of the most pioneering and influential works related to the field of bulk, thin film, and nanofiber energy harvesting devices and sensors. A brief history of piezoelectricity and its mechanism was provided followed by materials and technologies. Recent advancements in piezoelectric nanofiber energy harvesters and nanowire energy harvesters were also briefly reviewed.

Acknowledgments This research is supported by the National Research Foundation (NRF) Singapore under its Campus for Research Excellence and Technological Enterprise program. The Center for Environmental Sensing and Modeling (CENSAM) is an interdisciplinary research group of the Singapore-MIT Alliance for Research and Technology (SMART).

References

1. Ballas, R. G. (2007). *Piezoelectric multilayer beam bending actuators: Static and dynamic behavior and aspects of sensor integration.* New York: Springer Science & Business Media.
2. Damjanovic, D. (2008). Lead-based piezoelectric materials. In *Piezoelectric and acoustic materials for transducer applications* (pp. 59–79). Boston: Springer.

3. Panda, P. K. (2009). Review: Environmental friendly lead-free piezoelectric materials. *Journal of Materials Science, 44*, 5049–5062. https://doi.org/10.1007/s10853-009-3643-0.
4. Ramadan, K. S., Sameoto, D., & Evoy, S. (2014). A review of piezoelectric polymers as functional materials for electromechanical transducers. *Smart Materials and Structures, 23*, 033001. https://doi.org/10.1088/0964-1726/23/3/033001.
5. Sengupta, D., Kottapalli, A. G. P., Chen, S. H., et al. (2017). Characterization of single polyvinylidene fluoride (PVDF) nanofiber for flow sensing applications. *AIP Advances, 7*, 105205. https://doi.org/10.1063/1.4994968.
6. Sengupta, D., Kottapalli, A. G. P., Miao, J., & Kwok, C. Y. (2017). Electrospun polyvinylidene fluoride nanofiber mats for self-powered sensors. In *2017 IEEE SENSORS* (pp. 1–3). IEEE.
7. Tan, C. W., Kottapalli, A. G. P., Wang, Z. H., et al. (2011). Damping characteristics of a micromachined piezoelectric diaphragm-based pressure sensor for underwater applications. In *2011 16th international solid-state sensors, actuators and microsystems conference* (pp. 72–75). IEEE.
8. Asadnia, M., Kottapalli, A. G. P., Miao, J. M., & Triantafyllou, M. S. (2015). Ultra-sensitive and stretchable strain sensor based on piezoelectric polymeric nanofibers. In *Proceedings of the IEEE international conference on Micro Electro Mechanical Systems (MEMS)* (pp. 678–681).
9. Wang, Y. R., Zheng, J. M., Ren, G. Y., et al. (2011). A flexible piezoelectric force sensor based on PVDF fabrics. *Smart Materials and Structures, 20*. https://doi.org/10.1088/0964-1726/20/4/045009.
10. Shintaku, H., Nakagawa, T., Kitagawa, D., et al. (2010). Development of piezoelectric acoustic sensor with frequency selectivity for artificial cochlea. *Sensors and Actuators A: Physical, 158*, 183–192. https://doi.org/10.1016/j.sna.2009.12.021.
11. Li, C., Wu, P. M., Lee, S., et al. (2008). Flexible dome and bump shape piezoelectric tactile sensors using PVDF-TrFE copolymer. *Journal of Microelectromechanical Systems, 17*, 334–341. https://doi.org/10.1109/JMEMS.2007.911375.
12. Bora, M., Kottapalli, A. G. P., Miao, J. M., & Triantafyllou, M. S. (2017). Fish-inspired self-powered microelectromechanical flow sensor with biomimetic hydrogel cupula. *APL Materials, 5*. https://doi.org/10.1063/1.5009128.
13. Kottapalli, A. G. P., Asadnia, M., Miao, J. M., et al. (2012). A flexible liquid crystal polymer MEMS pressure sensor array for fish-like underwater sensing. *Smart Materials and Structures, 21*. https://doi.org/10.1088/0964-1726/21/11/115030.
14. Kottapalli, A. G. P., Tan, C. W., Olfatnia, M., et al. (2011). A liquid crystal polymer membrane MEMS sensor for flow rate and flow direction sensing applications. *Journal of Micromechanics and Microengineering, 21*. https://doi.org/10.1088/0960-1317/21/8/085006.
15. Lynch, J. P., Partridge, A., Law, K. H., et al. (2003). Design of piezoresistive MEMS-based accelerometer for integration with wireless sensing unit for structural monitoring. *Journal of Aerospace Engineering*. https://doi.org/10.1061/(ASCE)0893-1321(2003)16:3(108.
16. Wisitsoraat, A., Patthanasetakul, V., Lomas, T., & Tuantranont, A. (2007). Low cost thin film based piezoresistive MEMS tactile sensor. *Sensors and Actuators A: Physical*. https://doi.org/10.1016/j.sna.2006.10.037.
17. Thuau, D., Ayela, C., Poulin, P., & Dufour, I. (2014). Highly piezoresistive hybrid MEMS sensors. *Sensors and Actuators A: Physical*. https://doi.org/10.1016/j.sna.2014.01.037.
18. Mohammed, A. A. S., Moussa, W. A., & Lou, E. (2011). High-performance piezoresistive MEMS strain sensor with low thermal sensitivity. *Sensors*. https://doi.org/10.3390/s110201819.
19. Cao, L., Kim, T. S., Mantell, S. C., & Polla, D. L. (2000). Simulation and fabrication of piezoresistive membrane type MEMS strain sensors. *Sensors and Actuators A: Physical*. https://doi.org/10.1016/S0924-4247(99)00343-X.
20. Asadnia, M., Kottapalli, A. G. P., Karavitaki, K. D., et al. (2016). From biological cilia to artificial flow sensors: Biomimetic soft polymer nanosensors with high sensing performance. *Scientific Reports, 6*. https://doi.org/10.1038/srep32955.

21. Bora, M., Kottapalli, A. G. P., Miao, J., & Triantafyllou, M. S. (2017). Biomimetic hydrogel-CNT network induced enhancement of fluid-structure interactions for ultrasensitive nanosensors. *NPG Asia Materials, 9*, e440. https://doi.org/10.1038/am.2017.183.

22. Shapiro, Y., Kosa, G., & Wolf, A. (2014). Shape tracking of planar hyper-flexible beams via embedded PVDF deflection sensors. *IEEE/ASME Transactions on Mechatronics*. https://doi.org/10.1109/TMECH.2013.2278251.

23. Hansen, B. J., Liu, Y., Yang, R., & Wang, Z. L. (2010). Hybrid nanogenerator for concurrently harvesting biomechanical and biochemical energy. *ACS Nano, 4*, 3647–3652. https://doi.org/10.1021/nn100845b.

24. Zheng, J., He, A., Li, J., & Han, C. C. (2007). Polymorphism control of poly(vinylidene fluoride) through electrospinning. *Macromolecular Rapid Communications, 28*, 2159–2162. https://doi.org/10.1002/marc.200700544.

25. Pu, J., Yan, X., Jiang, Y., et al. (2010). Piezoelectric actuation of direct-write electrospun fibers. *Sensors and Actuators A: Physical, 164*, 131–136. https://doi.org/10.1016/j.sna.2010.09.019.

26. Liu, Z. H., Pan, C. T., Lin, L. W., et al. (2014). Direct-write PVDF nonwoven fiber fabric energy harvesters via the hollow cylindrical near-field electrospinning process. *Smart Materials and Structures, 23*. https://doi.org/10.1088/0964-1726/23/2/025003.

27. Chang, J., Dommer, M., Chang, C., & Lin, L. (2012). Piezoelectric nanofibers for energy scavenging applications. *Nano Energy, 1*, 356–371.

28. Harrison, J. S., & Ounaies, Z. (2002). Piezoelectricity and related properties of polymer films. In *Encyclopedia of polymer science and technology*. American Cancer Society.

29. Jean-Mistral, C., Basrour, S., & Chaillout, J.-J. (2010). Comparison of electroactive polymers for energy scavenging applications. *Smart Materials and Structures, 19*, 085012. https://doi.org/10.1088/0964-1726/19/8/085012.

30. Kim, J. Y. H., Cheng, A., & Tai, Y. C. (2011). Parylene-C as a piezoelectric material. In *Proceedings of the IEEE International Conference on Micro Electro Mechanical Systems (MEMS)* (pp. 473–476). Cacun, Mexico: IEEE.

31. Kim, J. Y. H., Nandra, M., & Tai, Y. C. (2012). Cantilever actuated by piezoelectric Parylene-C. *Proceedings of the IEEE International Conference on Micro Electro Mechanical Systems (MEMS).*, 1141–1144. Paris, France: IEEE.

32. Park, C., Ounaies, Z., Wise, K. E., & Harrison, J. S. (2004). In situ poling and imidization of amorphous piezoelectric polyimides. *Polymer (Guildf), 45*, 5417–5425. https://doi.org/10.1016/j.polymer.2004.05.057.

33. Newnham, R. E., Skinner, D. P., & Cross, L. E. (1978). Connectivity and piezoelectric-pyroelectric composites. *Materials Research Bulletin, 13*, 525–536. https://doi.org/10.1016/0025-5408(78)90161-7.

34. Pilgrim, S. M., Newnham, R. E., & Rohlfing, L. L. (1987). An extension of the composite nomenclature scheme. *Materials Research Bulletin, 22*, 677–684. https://doi.org/10.1016/0025-5408(87)90117-6.

35. Safari, A., Janas, V. F., & Bandyopadhyay, A. (1997). Development of fine-scale piezoelectric composites for transducers. *AICHE Journal, 43*, 2849–2856. https://doi.org/10.1002/aic.690431334.

36. 1–3 Composites. https://www.smart-material.com/13CompOverview.html. Accessed 23 Oct 2017.

37. Sessler, G. M., & West, J. E. (1962). Self-biased condenser microphone with high capacitance. *The Journal of the Acoustical Society of America, 34*, 1787–1788. https://doi.org/10.1121/1.1909130.

38. Gerhard-Multhaupt, R. (2002). Less can be more holes in polymers lead to a new paradigm of piezoelectric materials for electret transducers. *IEEE Transactions on Dielectrics and Electrical Insulation, 9*, 850–859. https://doi.org/10.1109/TDEI.2002.1038668.

39. Anton, S. R., & Sodano, H. A. (2007). A review of power harvesting using piezoelectric materials (2003–2006). *Smart Materials and Structures, 16*, R1–R21. https://doi.org/10.1088/0964-1726/16/3/R01.

40. Gonzalez, L., & Rubio a, M. F. (2001). A prospect on the use of piezoelectric effect to supply power to wearable electronic devices. In *Int Conf Intell robot Syst* (pp. 202–207). IEEE.
41. Niu, P., Chapman, P., Riemer, R., & Zhang, X. (2004). Evaluation of motions and actuation methods for biomechanical energy harvesting. *PESC Record – IEEE Annual Power Electronics Specialists Conference*, 2100–2106. Aachen, Germany: IEEE.
42. Beeby, S. P., Tudor, M. J., & White, N. M. (2006). Energy harvesting vibration sources for microsystems applications. *Measurement Science and Technology, 17*, R175–R195. https://doi.org/10.1088/0957-0233/17/12/R01.
43. Starner, T. (1996). Human-powered wearable computing. *IBM Systems Journal, 35*, 618–629. https://doi.org/10.1147/sj.353.0618.
44. Kymissis, J., Kendall, C., Paradiso, J., & Gershenfeld, N. (1998). Parasitic power harvesting in shoes. In *Dig Pap Second Int Symp Wearable Comput (Cat No98EX215)* (pp. 2–9). https://doi.org/10.1109/ISWC.1998.729539.
45. Shenck, N. S., & Paradiso, J. A. (2001). Energy scavenging with shoe-mounted piezoelectrics. *IEEE Micro, 21*, 30–42. https://doi.org/10.1109/40.928763.
46. Hellbaum, R. F., Bryant, R. G., Fox, R. L., Jalink, Jr. A. (1997). Thin layer composite unimorph ferroelectric driver and sensor. U.S. Pat. 5,632,841.
47. Sohn, J. W., Choi, S. B., & Lee, D. Y. (2005). An investigation on piezoelectric energy harvesting for MEMS power sources. *Proceedings of the Institution of Mechanical Engineers, Part C: Journal of Mechanical Engineering Science, 219*, 429–436. https://doi.org/10.1243/0954406 05X16947.
48. Platt, S. R. P. S. R., Farritor, S. F. S., & Haider, H. H. H. (2005). On low-frequency electric power generation with PZT ceramics. *IEEE/ASME Transactions on Mechatronics, 10*, 240–252. https://doi.org/10.1109/TMECH.2005.844704.
49. Roundy, S., & Wright, P. K. (2004). A piezoelectric vibration based generator for wireless electronics. *Smart Materials and Structures, 13*, 1131–1142. https://doi.org/10.1088/0964-1726/13/5/018.
50. Jeon, Y. B., Sood, R., Jeong, J. H., & Kim, S. G. (2005). MEMS power generator with transverse mode thin film PZT. *Sensors and Actuators, A: Physical, 122*, 16–22. https://doi.org/10.1016/j.sna.2004.12.032.
51. Lee, B. S., Lin, S. C., Wu, W. J., et al. (2009). Piezoelectric MEMS generators fabricated with an aerosol deposition PZT thin film. *Journal of Micromechanics and Microengineering, 19*, 065014. https://doi.org/10.1088/0960-1317/19/6/065014.
52. Wang, X. Y., Lee, C. Y., Peng, C. J., et al. (2008). A micrometer scale and low temperature PZT thick film MEMS process utilizing an aerosol deposition method. *Sensors and Actuators, A: Physical, 143*, 469–474. https://doi.org/10.1016/j.sna.2007.11.027.
53. Wang, X.-Y., Lee, C.-Y., Hu, Y.-C., et al. (2008). The fabrication of silicon-based PZT microstructures using an aerosol deposition method. *Journal of Micromechanics and Microengineering, 18*, 055034-1–055034-7. https://doi.org/10.1088/0960-1317/18/5/055034.
54. Aktakka, E. E., Peterson, R. L., & Najafi, K. (2010). A CMOS-compatible piezoelectric vibration energy scavenger based on the integration of bulk PZT films on silicon. In *Technical digest – international electron devices meeting, IEDM*. San Francisco, USA: IEEE.
55. Ataur Rahman, M., Lee, B.-C., Phan, D.-T., & Chung, G.-S. (2013). Fabrication and characterization of highly efficient flexible energy harvesters using PVDF–graphene nanocomposites. *Smart Materials and Structures, 22*, 085017. https://doi.org/10.1088/0964-1726/22/8/085017.
56. Won, S. S., Sheldon, M., Mostovych, N., et al. (2015). Piezoelectric poly(vinylidene fluoride trifluoroethylene) thin film-based power generators using paper substrates for wearable device applications. *Applied Physics Letters, 107*. https://doi.org/10.1063/1.4935557.
57. Huang, Z. M., Zhang, Y. Z., Kotaki, M., & Ramakrishna, S. (2003). A review on polymer nanofibers by electrospinning and their applications in nanocomposites. *Composites Science and Technology, 63*, 2223–2253. https://doi.org/10.1016/S0266-3538(03)00178-7.
58. Reneker, D. H., Yarin, A. L., Zussman, E., & Xu, H. (2007). Electrospinning of nanofibers from polymer solutions and melts. *Advances in Applied Mechanics, 41*, 43–346.

59. Li, D., & Xia, Y. (2004). Electrospinning of nanofibers: Reinventing the wheel? *Advanced Materials, 16*, 1151–1170.
60. Subbiah, T., Bhat, G. S., Tock, R. W., et al. (2005). Electrospinning of nanofibers. *Journal of Applied Polymer Science, 96*, 557–569. https://doi.org/10.1002/app.21481.
61. Pham, Q. P., Sharma, U., & Mikos, A. G. (2006). Electrospinning of polymeric nanofibers for tissue engineering applications: A review. *Tissue Engineering, 12*, 1197–1211. https://doi.org/10.1089/ten.2006.12.1197.
62. Chang, J., Dommer, M., Chang, C., & Lin, L. (2012). Piezoelectric nanofibers for energy scavenging applications. *Nano Energy, 1*, 356–371.
63. Sun, D., Chang, C., Li, S., & Lin, L. (2006). Near-field electrospinning. *Nano Letters, 6*, 839–842. https://doi.org/10.1021/nl0602701.
64. Chen, X., Xu, S., Yao, N., & Shi, Y. (2010). 1.6 v nanogenerator for mechanical energy harvesting using PZT nanofibers. *Nano Letters, 10*, 2133–2137. https://doi.org/10.1021/nl100812k.
65. Chen, X., Xu, S., Yao, N., et al. (2009). Potential measurement from a single lead ziroconate titanate nanofiber using a nanomanipulator. *Applied Physics Letters, 94*. https://doi.org/10.1063/1.3157837.
66. Chang, C., Tran, V. H., Wang, J., et al. (2010). Direct-write piezoelectric polymeric nanogenerator with high energy conversion efficiency. *Nano Letters, 10*, 726–731. https://doi.org/10.1021/nl9040719.
67. Sencadas, V., Gregorio, R., & Lanceros-Méndez, S. (2009). α to β phase transformation and microstructural changes of PVDF films induced by uniaxial stretch. *Journal of Macromolecular Science, Part B Physics, 48*, 514–525. https://doi.org/10.1080/00222340902837527.
68. Li, D., Wang, Y., Xia, Y., et al. (2004). Electrospinning nanofibers as Uniaxially aligned arrays and layer-by-layer stacked films. *Advanced Materials, 16*, 361–366. https://doi.org/10.1002/adma.200306226.
69. Mandal, D., Yoon, S., & Kim, K. J. (2011). Origin of piezoelectricity in an electrospun poly(vinylidene fluoride-trifluoroethylene) nanofiber web-based nanogenerator and nano-pressure sensor. *Macromolecular Rapid Communications, 32*, 831–837. https://doi.org/10.1002/marc.201100040.
70. Liu, Z. H., Pan, C. T., Lin, L. W., et al. (2014). Direct-write PVDF nonwoven fiber fabric energy harvesters via the hollow cylindrical near-field electrospinning process. *Smart Materials and Structures, 23*, 025003. https://doi.org/10.1088/0964-1726/23/2/025003.
71. Xu, S., Hansen, B. J., & Wang, Z. L. (2010). Piezoelectric-nanowire-enabled power source for driving wireless microelectronics. *Nature Communications, 1*, 93. https://doi.org/10.1038/ncomms1098.
72. Koka A, Sodano HA. (2014) A low-frequency energy harvester from ultralong, vertically aligned BaTiO 3 nanowire arrays. Advanced Energy Materials n/a-n/a. https://doi.org/10.1002/aenm.201301660.
73. Wang, Z. L., & Song, J. (2006). Piezoelectric nanogenerators based on zinc oxide nanowire arrays. *Science (80-), 312*, 242–246. https://doi.org/10.1126/science.1124005.
74. Wang, X., Song, J., Liu, J., & Wang, Z. L. (2007). Direct-current nanogenerator driven by ultrasonic waves. *Science (80-), 316*, 102–105. https://doi.org/10.1126/science.1139366.
75. Qin, Y., Qin, Y., Wang, X., et al. (2008). Microfibre-nanowire hybrid structure for energy scavenging. *Nature, 451*, 809–813. https://doi.org/10.1038/nature06601.

Chapter 3
Nature-Inspired Self-Powered Sensors and Energy Harvesters

Debarun Sengupta, Ssu-Han Chen, and Ajay Giri Prakash Kottapalli

3.1 Introduction

Biomimetic and nature-inspired design philosophies have been gaining popularity among modern researchers for past few decades. During product development, efficient design is the key to enhancing the performance of the device. For example, while designing a locomotive, a key concern is to have an overall aerodynamic architecture to reduce air resistance. There are many other examples we might find in daily life where design might be of utmost importance. Living creatures have evolved over the ages through the process of natural selection to adapt to their surrounding environment. Taking inspiration from living beings to solve engineering problems is often the best design approach as it saves a lot of time in design optimization. For example, the sophisticated sensing capabilities demonstrated by the aquatic creatures make use of complex biological sensors having multilayered functionalities.

D. Sengupta
Department of Advanced Production Engineering, Engineering and Technology Institute Groningen (ENTEG), Faculty of Science and Engineering, University of Groningen, Groningen, The Netherlands

S.-H. Chen
Microsystems Research, School of Electrical Engineering and Telecommunications, The University of New South Wales, Kensington, Australia

A. G. P. Kottapalli (✉)
Department of Advanced Production Engineering, Engineering and Technology Institute Groningen (ENTEG), Faculty of Science and Engineering, University of Groningen, Groningen, The Netherlands

MIT Sea Grant College Programme, Massachusetts Institute of Technology (MIT), Cambridge, MA, USA
e-mail: a.g.p.kottapalli@rug.nl

© The Author(s), under exclusive licence to Springer Nature Switzerland AG 2019
A. G. P. Kottapalli et al., *Self-Powered and Soft Polymer MEMS/NEMS Devices*,
SpringerBriefs in Applied Sciences and Technology,
https://doi.org/10.1007/978-3-030-05554-7_3

With the rapid growth of semiconductor industry for past two decades, there has been tremendous growth in the field of semiconductor sensor technologies. Development of micro/nano-fabrication techniques has led to the increase of micro/nano-system-based sensors like pressure sensors, heat sensors, and flow sensors.

With the demand in market for wearable and other consumer electronics devices, the focus of the industry is now to develop state-of-the-art sensors capable of carrying out sensing tasks at low-power budget. This is where the philosophy of bioinspiration can play an essential role by virtue of having nature evolved clever design concept tailored to a sensing task.

The sensing capabilities demonstrated by underwater creatures have drawn the interests of engineers and material scientists for last three decades. Significant research efforts have been put in to understand the sensing principles employed by aquatic creatures to navigate in turbid underwater environments. Design and development of efficient self-powered sensors are the primary motivation behind understanding the sensing principles, morphologies, and functionalities of various natural sensors found in underwater animals.

3.1.1 Flow Sensing in Marine Creatures

Evolution has led to the development of some of the most impressive flow sensing capabilities in underwater creatures. For surviving in harsh underwater environments full of predators, fishes have evolved with sensors capable of sensing flows accurately. Many marine animals are capable of detecting prey and predators by forming a three-dimensional map of their surrounding by sensing velocity and pressure field surrounding them [1]. This is achieved by employing an array of flow sensors distributed along their bodies.

For example, blind cave fishes living in dark caves lack vision and yet are capable of generating a three-dimensional map of their habitat by sensing pressure and velocity variation in their immediate surroundings [2, 3]. This sophisticated hydrodynamic sensing is achieved by employing an array of sensors also known as neuromasts embedded into their lateral lines.

Crocodiles are known for employing dome-shaped pressure receptors which are capable of sensing surface waves generated due to the tiniest drop of water. By doing so crocodiles can orient themselves to surface waves in complete darkness [4].

Another interesting example is of the harbor seals utilizing their whiskers to sense wake signatures left by the prey up to 30 s after it has already passed. Their whiskers are sensitive enough to detect a flow having a magnitude as low as 245 μm s^{-1} [5].

3.1.2 Flow Sensing in Fishes

The lateral line in a fish consists of an array of sensors known as neuromasts. They are pressure-gradient sensors capable of sensing slightest perturbance in the flow fields surrounding the fish due to the presence of any obstacles or objects. Fishes

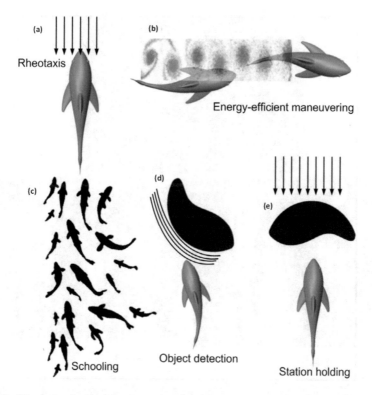

Fig. 3.1 Schematic diagram showing various lateral-line mediated behavior demonstrated by fishes. (Figure reproduced from [12] with permission of ©IOPSCIENCE)

demonstrate a range of behavior mediated by lateral-line including rheotaxis, energy-efficient maneuvering, schooling, object detection, and station holding [6–11]. Figure 3.1 is a schematic representation of some of the interesting lateral-line mediated behavior observed in fishes.

Lateral-line systems (LLS) are of utmost importance in blind cave fishes (Fig. 3.2a) as they depend on them for sensing the surroundings. Due to their functional significance, the LLS in fishes are sometimes referred to as "touch at a distance" [13]. As stated earlier, LLS comprise of the smallest sensory unit known as neuromasts which in turn consists of hair cells, stereocilia bundles, kinocilium, and nerves. The neuromast with stereocilia bundles is encapsulated by an elongated or dome-shaped structure made of a soft porous gel-like material called as cupula.

Depending on their location, neuromasts can be classified into two categories:

- Superficial neuromasts (SNs): This type of neuromast is present on the surface of the fish skin.
- Canal neuromasts (CNs): This type of neuromast is present below the surface of the skin embedded in fluid-filled canals (Fig. 3.2c). The canal opens up on both ends, to the skin water interface by means of pores.

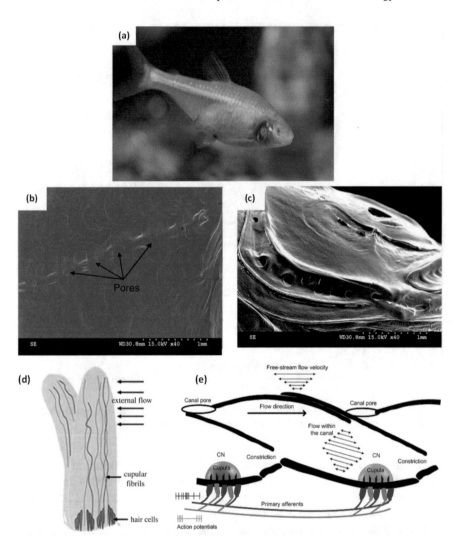

Fig. 3.2 (**a**) Image of a blind cave fish [17]. (**b**) SEM micrograph showing the lateral line found in cave fishes. (Figure reproduced from [18] with permission of ©Sagepub). (**c**) SEM images showing the canal pores of CNs [19]. (**d**) Schematic diagram showing the structure of an SN and explaining its working principle [19]. (**e**) Schematic diagram showing the structure of a CN and explaining its working principle [19]

The main differences between SNs and CNs lie in the structure of their cupula and stereocilia. SNs have an elongated cupula with an elongated tip (Fig. 3.2d), whereas the cupulae found in CNs are dome-shaped (Fig. 3.2e). The elastic modulus of cupula of SN (10 Pa) is significantly lower than that of cupula of CN having an elastic modulus of 10 kPa [14–16]. Both CNs and SNs are characterized by two spatially intermingled and oppositely oriented groups of hair cells.

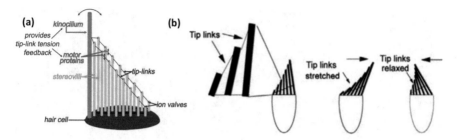

Fig. 3.3 (**a**) Schematic representation of hair cell structure comprising kinocilium and stereocilia. (Figure reproduced from [22] with permission of WILEY-VCH.) (**b**) Schematic representation of stretching and relaxing of tip links in stereocilia leading to excitatory and inhibitory response of neuromasts. (Figure reproduced from [23] with permission of Elsevier Inc.)

The sensing mechanism of SN is fundamentally different from CN. The water which flows past the skin of the fish interacts with the SNs generating responses proportional to the velocity of the water flow. In case of the CNs, the water entering a canal through the skin pores interacts with the sensors generating responses which are proportional to the net acceleration of the water flow [20, 21].

Figure 3.3a shows the schematic diagram of hair cell having stereocilia and kinocilium. Water interacting with the cupula deflects it which starts the transduction process involving inhibition or excitation through mechanically sensitive ion channel. Tip links connect the stereocilia at their tips (Fig. 3.3a). Figure 3.3b presents the bending of stereocilia toward or away from kinocilium causing stretching/relaxation of the tip links. When the stereocilia are bent, the tip links are either stretched or relaxed depending on the direction of the bending of the stereocilia. When stereocilia bend toward the long kinocilium, the tip links are stretched leading to the initiation of the transport of ions across the hair cell membrane [23, 24]. Bending of stereocilia in the opposite direction leads to relaxation of the tip links causing prevention of ion transport. The responses in intermediate directions are cosine functions of input directions.

Due to the structural differences in SNs and CNs, they respond differently to different frequency domains. SNs are like low-pass filters which react to low-frequency stimuli, whereas the CNs behave like high-pass filters responding to stimuli in the higher-frequency domain [11, 23]. Combining both SNs and CNs, fishes can filter out low-frequency noises from high-frequency large amplitude signals, thus improving the overall signal-to-noise ratio.

3.1.3 Flow Sensing in Crocodiles

Crocodiles achieve passive sensing through mechanoreceptors called integumentary sensory organs (ISOs) [25]; they are sometimes known as dome pressure receptors (DPRs) [4]. These receptor organs are densely populated on the jaws and the

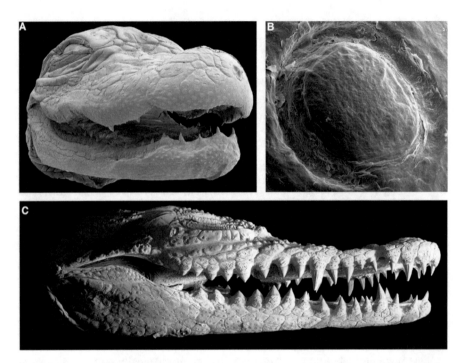

Fig. 3.4 Cranial region of a crocodilian viewed under the scanning electron microscope with false colorization: (**a**) the cranial region of an alligator with ISOs (yellow) shown, (**b**) a single ISO of an alligator shown at higher magnification, (**c**) the cranial region of a crocodile with distributed ISO shown (yellow). (Figure reproduced from [25] with permission of Journal of Experimental Biology)

rest of the body of the crocodile with dome-shaped structures varying from 0.2 to 1.2 mm in diameter [25], as shown in Fig. 3.4a–c. The ISOs consist of slowly adapting (SA) and rapidly adapting (RA) receptors which enable crocodiles to locate the source of disturbances both on the surface of the water and underwater to hunt for preys in dark and murky environments. These mechanoreceptors are categorized into two types of receptor cells – (i) SA receptors are associated with Merkel cells [25] which respond to constant intensity level of stimulus; (ii) RA receptors are related to lamellated corpuscles [25] which react to a change in the intensity level of the stimulus.

The response of the SA receptors to a square wave stimulus and that of RA receptors to a sinusoidal stimulus has been demonstrated by Leitch et al. [23]. From the response of the two stimulus signals, it can be verified that the SA receptors responded throughout the presence of the steady intensity level of the stimulus and the RA receptors responded to the same frequency as that of the sinusoidal stimulus. With these distinct response characteristics, the SA receptors are well suited for sensing steady flows or variations, whereas RA receptors are more suitable for sensing oscillatory pressures caused by fast-moving aquatic animals.

3.1.4 Flow Sensing in Harbor Seals

Several species of marine mammals have evolved in nature with highly efficient hydrodynamic sensory systems. For example, harbor seals use their whiskers (vibrissae) to track prey, evade predators, and identify conspecifics (Fig. 3.5a). In the natural habitat, it is crucial for the seals to have this ability to track the hydrodynamic trail of the fish to capture the swimming directions of their prey. Dehnhardt et al. have demonstrated a "go" and "no-go" testing paradigm on a trained blindfolded and acoustically masked harbor seal (*Phoca vitulina*) that restricted the seal to utilize its whiskers to detect flows. They identified that relying on whiskers, the seal is capable of detecting subtle water movements as low as 245 μm/s in the 10–100 Hz range [26]. When the cross-sectional shape of the whiskers has a close resemblance to a bluff body (cylindrical shape), the wake forms a double array of staggered vortices, known as the Kármán street. The trembling forces as a result of Kármán street act on the body with an amplitude comparable to its cross-sectional dimension and with high frequency. Such vibration also referred to as vortex-induced vibration (VIV) gives rise to oscillations in the cross-flow direction in which the corresponding flow profiles appear unsteady [27]. Remarkably, seal vibrissae are sensitive enough to detect minute changes in the flow left by marine animals while rejecting this self-induced flow noise. This noteworthy feat gives credit to the uniquely shaped whiskers with a variable cross section in the shape of an ellipsoid (Fig. 3.5b) [28]. In the spanwise direction, the geometry of the whisker varies sinusoidally, while the upstream undulation is in out of phase with the downstream undulation. With the specialized undulatory morphology, it is believed that the self-induced noise from VIV is suppressed due to this factor. These undulations are understood to disrupt the formation of Kármán vortex and cause the formation of streamwise vorticity, hence reducing substantially the fluid forces on the whisker [29]. At steady velocity, it is believed this feature allows the seal to move forward

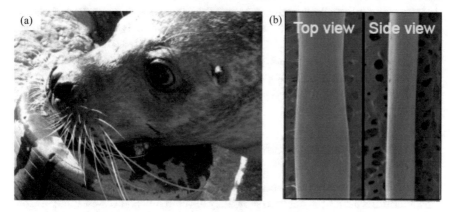

Fig. 3.5 (**a**) A harbor seal displaying its vibrissae, (**b**) SEM images of a whisker showing its undulatory, elliptical geometry. (Figure reproduced from [28] with permission of IEEE)

with minimal self-induced noise from VIV. However, seals following the trails of fish experience a more complex phenomenon. The body and caudal fin from a fish generate coherent ring-link vortical structures that induce a jet flow, as well as a drag wake, and it is only in the very far field when the two cancel each other for a self-propelled body. These wakes show predominance flow similar to that of a Kármán street, but the vortices rotate in opposite direction. Hence, understanding the hydrodynamic properties of fish detection by harbor seal whiskers can be modeled by the interaction between the wake of an upstream cylinder and a seal whisker to provide principal mechanisms that enable the fine sensitivity of the whiskers [30]. This undulation feature of the whiskers has been adopted by some biomimetic engineers in creating enhanced sensors to detect flow disturbances, which will be discussed in Sect. 3.2.3.

3.2 Underwater Animal-Inspired Self-Powered Sensors

Marine creatures have developed highly efficient sensory systems for their survival in harsh marine habitat. Engineers have taken inspiration from nature in replicating mechanoreceptors for applications in autonomous underwater vehicles (AUVs) and efficient flow sensors. These nature-inspired sensors provide passive sensing techniques that are, in comparison, more efficient than active sensing strategies. Due to this reason, scientists and engineers have taken great initiative in understanding how the sensory organs of the marine creatures have help them to avoid predators and hunt for preys.

In this section, various sensing techniques inspired by marine creatures like fish, crocodiles, and harbor seals are presented. The neuromasts from fishes have been studied extensively in the past and have been mimicked for various sensing applications. Integumentary sensory organs (ISOs) found in crocodiles are excellent receptors to detect steady-state pressures and oscillatory pressures under water. These receptors are replicated by engineers with MEMS technology to perform the same functionalities. Vibrissae organs are integral to harbor seals to navigate through water and sense hydrodynamic information when they are visually or acoustically impaired. The whiskers of the harbor seal have the ability to detect low fluid velocities in low-frequency range. Engineers have adapted this concept in reproducing artificial whiskers to sense minute disturbances in water.

3.2.1 Neuromast-Inspired Biomimetic MEMS Sensors

The fish LLS has inspired researchers in recent years to develop biomimetic MEMS sensors. Efforts have been put in to separately develop SN- and CN-inspired sensors for different sensing applications. The most widely reported artificial hair cell-based flow sensor designs are characterized by high aspect-ratio pillar-like structures on

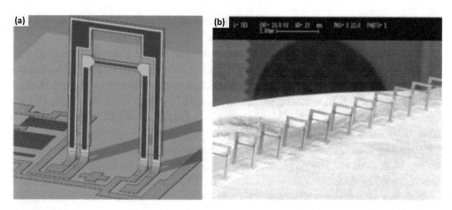

Fig. 3.6 (**a**) Schematic representation of individual microfabricated out-of-plane artificial hair cell sensors. (**b**) Scanning electron microscope image of artificial lateral line with 16 sensors spaced 1 mm apart. (Figure reproduced from [31], Copyright (2006) National Academy of Sciences, USA)

substrates known as the cilium. The structural configuration of such sensors allows for a densely packed array of sensors which pose minimum intrusion to the flow field. The critical concern while developing such sensors for the underwater environment is that they should be able to survive in the harsh marine environment irrespective of the fabrication method being employed to create such sensors.

In one of the early works, Yang et al. reported a hot wire anemometer (HWA) principle-based LLS comprising of a linear array of flow sensors, similar to the array of neuromasts found in fishes [25]. In this work, they used a unique microfabrication process combining an efficient three-dimensional assembly technique with traditional surface micromachining method to develop an array of novel hot-wire anemometer. The individual structure of each sensor consists of a 400 μm long hot wire element made of nickel filament sandwiched between two layers of polyimide which is elevated 600 μm out of the plane by two prongs (Fig. 3.6a). The nickel hot wire was reported to have a temperature coefficient of resistance of 4100 ppm/°C. The array of sensors consisted of 16 sensors with an intersensor spacing of 1 mm (Fig. 3.6b). For parallel data acquisition and noise-floor reduction, each sensor was integrated with monolithically integrated metal-oxide semiconductor circuitry. The testing of the array of sensors revealed a nonlinear characteristic in a water channel with flow velocities up to 0.25 m/s. This work demonstrated the potential of biomimetic flow sensing in an underwater environment. The authors were able to show the ability of their sensor in dipole source localization, hence discerning the dipolar near field generated by any nearby object.

Though this work was one of the firsts to show that artificial lateral line can successfully perform hydrodynamic wake detection and dipole source localization, the sensors demonstrated in the work differed from its real biological equivalent in many ways. The main difference between this artificial lateral line and real LLS is that the real LLS consist of both CNs and SNs, and each neuromast has a cupula encapsulating a bundle of hair cells, whereas the artificial lateral line only has a certain number of HWAs placed superficially.

In another work, Chen et al. developed biological hair cell-inspired artificial hair cell (AHC) sensor comprising of a high-aspect-ratio cilium structure attached at the distal end of a silicon cantilever beam [32]. The main sensing element in this AHC is a piezoresistive silicon strain gauge at the base of the cantilever. The cilium structure was made of SU-8 and was up to 700 μm in height. The sensor was reported to have high sensitivity and directionality. On subjecting to oscillatory (ac) flows generated with a dipole setup, the sensor was able to respond to velocity amplitudes as low as 0.7 mm/s. The angular resolution for the wind-tunnel directionality test was reported as 2.16°.

Building on previous works, Peleshenko et al. developed a hybrid design combining hard and soft materials to develop AHC sensor having cupula like encapsulating structure. The sensor design in the work comprised of a silicon cantilever membrane with 600 μm tall SU-8 pillar. To mimic the cupula, a viscous hydrogel synthesized by means of photopatterned polymerization of polyethylene glycol (PEG) was drop casted to form a dome shape over the SU-8 (Fig. 3.7a). The drop casting was followed by cross-linking and swelling in water at ambient conditions.

Both of the sensors with and without cupula were subjected to steady and oscillatory flows in underwater environment. The sensor with cupula demonstrated

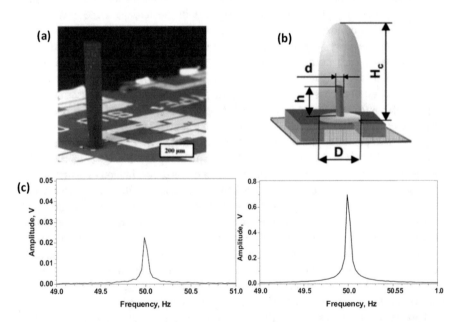

Fig. 3.7 (a) SEM image of the artificial hair cell-based flow sensors without hydrogel cupula. (b) Schematic diagram of the artificial hair cell sensor with hydrogel cupula. (c) Plot comparing the response of the naked hair cell sensor with the hydrogel-capped sensor subjected to 50 Hz oscillatory flow. (Figure reproduced from [14] with permission of WILEY-VCH Verlag GmbH and Co. KGaA, Weinheim)

10–30 times the signal intensity than the naked hair cell design. For a 50 Hz oscil-latory flow signal, the sensor with PEG hydrogel cupula demonstrated significantly better performance than its naked counterpart (Fig. 3.7c). Also, the hydrogel cupula improved the overall minimum velocity detection threshold by 2.5 times (from 0.2 mm s^{-1} to 0.075 mm s^{-1}).

Recently, Bora et al. reported the development of MEMS-based CN-inspired self-powered sensor and demonstrated its capability of indirectly detecting steady-state flow by employing the principle of vortex-induced vibrations (VIV) [33]. The sensor reported in this work consists of three distinct parts (Fig. 3.8a, b):

1. Piezoelectric sensing element: A membrane fabricated using lead zirconate tita-nate (PZT) was used as the main sensing element.
2. The hair cell structure: A high aspect ratio cylindrical pillar of copper (Cu) placed in the middle of the PZT membrane was used for mimicking the tall hair cell structure. This cu pillar interacts with the fluid flow and detects disturbances in the fluid.
3. Hydrogel cupula: To mimic the dome-shaped cupula of the CNs, hyaluronic acid (HA) modified with methacrylic anhydride (MA) was drop cast over the cu pillar.

Figure 3.9a, b shows the CN-inspired sensor with and without hydrogel cupula.

The main problem with all piezoelectric sensors is electron discharging when subjected to static forces. In this work, an innovative approach to solve the problem associated with electron discharging was proposed. Vibration of the PZT membrane caused by the VIV generated on the Cu pillar is used for determination of the flow

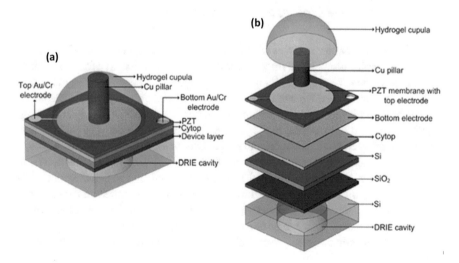

Fig. 3.8 (a) Schematic diagram of CN-inspired flow sensor with HA-MA hydrogel cupula. (b) Exploded schematic view of CN-inspired flow sensor with HA-MA hydrogel cupula. (Figure reproduced from [33] with permission of AIP)

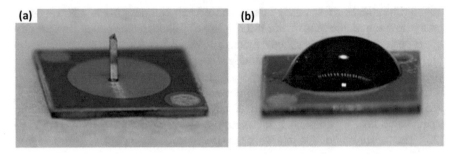

Fig. 3.9 (**a**) Photograph of the naked sensor. (**b**) Photograph of the sensor with cupula. (Figure reproduced from [33] with permission of AIP)

velocity (U). Steady-state flow sensing test was conducted on the sensor for a range of velocities (0–0.7 m/s). Figure 3.10a shows the sensor output for a steady-state flow velocity of 0.625 m/s. Oscillatory flow tests were performed on the sensor by using a vibrating sphere (dipole) to agitate the water surrounding the sensor. The dipole was driven by a series of sinusoidal signals in a plane perpendicular to the axis of the Cu pillar. The response of the sensor closely followed the driving frequency of the dipole (Fig. 3.10c). Experiments to demonstrate the advantage of having a hydrogel cupula as opposed to the naked pillar revealed a performance improvement of 2.1 times when subjected to an oscillatory flow test (at 35 Hz) (Fig. 3.10b).

3.2.2 Crocodile-Inspired Self-Powered Sensors

Autonomous underwater vehicles (AUVs) are integral instruments when it comes to underwater exploration, surveying, and military applications. In order to achieve autonomous maneuvers, the AUVs must be equipped with sensing elements to perceive their surrounding environments and enhance navigation control. Traditionally, this has been achieved by active sensing strategies like sound navigation and ranging (SONAR) and optical imaging [34]. Even though sonar technology is the most mature and adapted method for detecting the underwater environments for AUVs, it comes with inherent issues like sonar blind zones [35] and poses fatal threats to aquatic lives due to its intense sound waves transmitted by the sonar system. For optical imaging, the reliability of the environmental mapping relies on the visibility of the water [31]. Moreover, with active sensing strategies, the equipment accompanied with the sensors adds weight to the AUVs, making them energy inefficient. However, the aforementioned drawbacks can be overcome with passive sensing technologies.

By taking inspiration from the sensory organs of the crocodiles, implementing passive sensing in AUVs could potentially be possible by mimicking the ISOs of the crocodiles. Biomimetic engineers have attempted to recreate the SA and RA

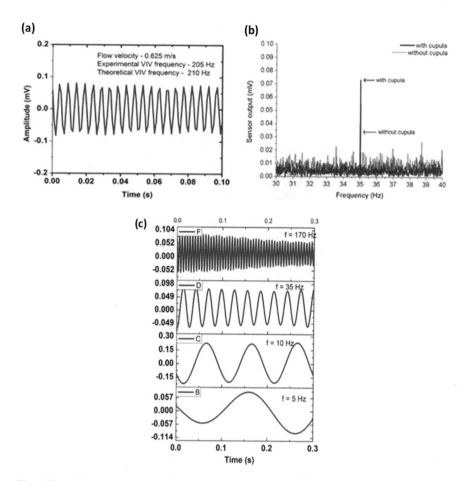

Fig. 3.10 (**a**) Response of the sensor for a steady-state flow velocity of 0.625 m/s. (**b**) Comparing the responses of the sensors with and without cupula. (**c**) Oscillatory flow responses of the sensors at various frequencies. (Figure reproduced from [33] with permission of AIP)

receptors with MEMS-based sensors [36]. The SA and RA dome-shaped receptors are constructed with piezoresistive sensors to sense steady pressures and piezo-electric sensors to sense oscillating pressures, respectively (Fig. 3.11a, b). The SA and RA domes consist of five MEMS piezoresistive pressure sensors and five MEMS piezoelectric pressure sensors, mounted on a 3D-printed polymer dome (Fig. 3.11c, d).

The replicas of the dome-shaped receptors of the ISOs on crocodiles are advantageous to sense information in three dimensions due to its spherically symmetric structure. From the patterns and combinations of the pressure value readings by the individual sensors, locating the originality of the disturbance can be determined. From experimental results by Kanhere et al. [36], the ability of SA and RA dome

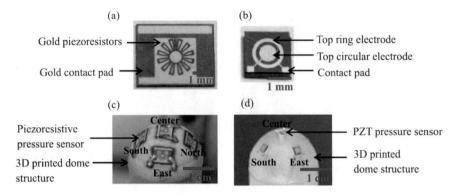

Fig. 3.11 Bioinspired MEMS SA and RA dome receptors: (**a**) piezoresistive gold sensor on LCP membrane, (**b**) micro-diaphragm piezoelectric pressure sensor, (**c**) SA dome with piezoresistive pressure sensors, (**d**) RA dome with piezoelectric pressure sensors. (Figure reproduced from [36] with permission of IOP publishing)

receptors to distinguish the direction of steady-state flows and oscillatory disturbances by determining the outputs of the five pressure sensors on each dome has been demonstrated.

3.2.3 Harbor Seal-Inspired Biomimetic Sensors

AUVs not only reply on sensing elements to perceive their surrounding environments but also determining fluid flow to aid in navigation is also crucial for autonomous control. Various commercial and military applications rely on fluid flow information to interpret target movements in water. Several researches have acquired their inspiration from harbor seals in developing highly sensitive artificial whiskers to detect hydrodynamic trails. The harbor seal can respond to extremely weak hydrodynamic stimuli with their whiskers as aforementioned. This has sparked many interests in mimicking whisker-like sensors to detect fluid motion for various applications [37–40].

A capacitive-based whisker-like sensor for fluid motion sensing has been demonstrated by Stocking et al. [37]. The sensor features a rigid artificial whisker mounted on a cone-in-cone parallel-plate capacitor base which are separated into four quadrants to provide directional information for the fluid flow. The gap between the electrodes is filled with silicone oil to obtain a dielectric constant of about 2.5. The damping and restoring force of the artificial whisker are provided by a 200 μm thick polydimethylsiloxane (PDMS) membrane, as shown in Fig. 3.12a, b. The sensor has demonstrated the ability to distinguish flow-induced forces for both steady and oscillatory conditions with millinewton-level precision. The output signal from each of the quadrants allowed the interpretation of flow direction. However, achiev-

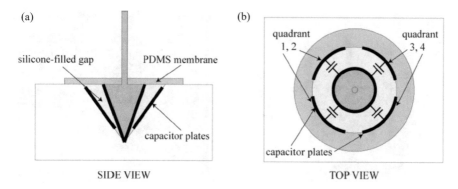

Fig. 3.12 Schematic drawings of the cone-in-cone sensor design: (**a**) side view of the sensor, (**b**) top view of the sensor. (Figure reproduced from [37] with permission of IEEE)

ing a similar sensitivity among the quadrants is particularly challenging with this design due to the requirement for the inner cone to be perfectly aligned to the outer cone. The same group later presented a Wake Information Detection and Tracking System (WIDTS) with an array of eight capacitive-based whisker-like sensors developed from their previous work [41]. The individual sensors in the WIDTS are designed to measure the direction and speed of fluid motion. As a proof of concept, the WIDTS has been fitted to a trained harbor seal via a bite plate to hold securely in the mouth of the seal. The WIDTS is then exposed to a hydrodynamic path left by a miniaturized submarine guided by the tracking behavior of the seal. This provided a realistic sensing environment for the WIDTS to detect underwater disturbances similar to the environment encountered by the seal. From observing the tracking path of the seal, the WIDTS also demonstrated detection of the hydrodynamic disturbances triggered by the variations of the seal's tracking behavior. This provided evidence that the artificial whisker arrays may achieve underwater sensing capabilities.

Alternative whisker-like sensors have been proposed by Alvarado et al. [39, 40]. The design has a more close resemblance to that of a whisker follicle sinus complex system (FSC). Figure 3.13a illustrates the major structural components inside the follicle sinus complex and a simple mechanical model of the whisker-follicle system. Figure 3.13b outlines the basic whisker sensor design. The whisker with length $L+\ell$ is supported by a viscoelastic membrane which mimics the FSC. Inside the sensor capsule are four flexible displacement sensors to detect whisker base oscillations θ in two perpendicular plans of motion (along and across the direction of motion). The flexible sensors consist of commercially available piezoresistive-based sensors, Bend Sensor®, that changes its electrical conductivity as it is bent (http://www.flexpoint. com/). The viscoelastic membranes used are silicone-based rubbers with mechanical properties close (Young's modulus and viscosity) to that of FSC tissues (Ecoflex 0010, Ecoflex 0030, and MoldMax 30) (http://www.smooth-on.com).

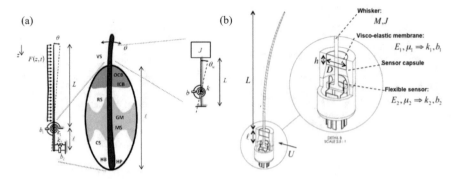

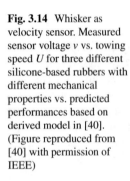

Fig. 3.13 (**a**) Whisker-FSC system and a simple lumped parameter model. (**b**) Basic whisker-like sensor design. (Figure reproduced from [40] with permission of IEEE)

Fig. 3.14 Whisker as velocity sensor. Measured sensor voltage v vs. towing speed U for three different silicone-based rubbers with different mechanical properties vs. predicted performances based on derived model in [40]. (Figure reproduced from [40] with permission of IEEE)

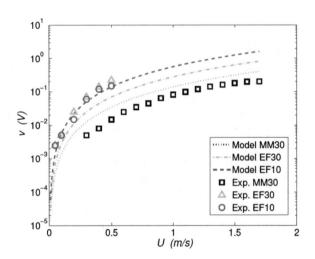

   The whisker-like sensors have been tested in water by fixing the whisker modules on a carriage and towed at constant speeds along a 10 m × 2 m × 1.5 m water tank. The sensors are capable of detecting flow speeds spanning from 0.05 m/s up to 2 m/s with a whisker length L of 0.17 m, as shown in Fig. 3.14. With mindful design considerations, the artificial FSC can be tailored to sense a wide range of velocities for specific applications.

   The aforementioned whisker-like sensors employ cylindrical standing pillars or strands to mimic the whiskers of the harbor seals. When these artificial whiskers are subjected to flow disturbances, vortex-induced vibration other than drag pressure are induced by the steady flow. The whiskers in seals have developed unique geometry along the length of their whiskers which is believed to have an important role in suppressing vortex-induced vibrations [29]. More recently, bioinspired artificial

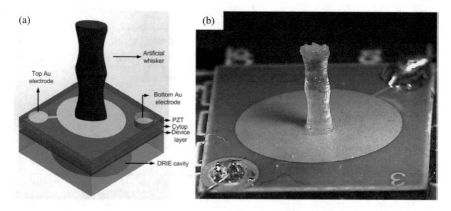

Fig. 3.15 (**a**) Schematic of the micro-whisker sensor (**b**) fabricated device with the artificial polymer micro-whisker fabricated by stereolithography mounted at the center of the PZT membrane. (Figure reproduced from [42] with permission of IEEE)

whisker fabricated by stereolithography (SLA) coupled with a piezoelectric MEMS sensing membrane has been demonstrated with the capability to suppress vortex-induced vibrations (Fig. 3.15a) [42, 43]. The micro-pillar mimicking the real seal whisker is fabricated from a UV curable Si60 polymer material with a high-resolution VIPER™ SLA® system. The varying spatial features are formed layer by layer with SLA processing techniques. The piezoelectric sensing diaphragm is formed by bonding a bulk PZT plate to an SOI wafer using spun-on Crypto as an intermediate layer. The PZT plate is flanked by two Cr/Au metal electrodes that form as the contact pads to sense the output piezoelectric voltages. The sensing membrane is released from the backside by DRIE. Figure 3.15b shows the sensing device with the micro-pillar mounted at the center of the PZT membrane.

The performance of the sensor is evaluated using a vibrating sphere stimulus in a water tank. The threshold and sensitivity of the sensor is tested by varying the amplitude of the sinusoidal signal supplied to the dipole. The output signal from the sensor is amplified by a gain of 500, and the resultant peak-to-peak signal amplitudes from various oscillatory flow velocities are illustrated in Fig. 3.16a. The sensor demonstrated a linear response up to an oscillatory flow velocity of 250 mm/s, with a lowest detectable velocity of 193 μm/s. The capability to suppress vortex-induced vibrations in low flow velocities due to the whisker-like undulations of the micro-pillar is demonstrated by comparing to a cylindrical micro-pillar in a steady-state flow. The two types of sensors are tested in a water tunnel with the water flow in perpendicular to the long axis of the micro-pillars. The vortex-induced vibrations for the cylindrical and undulated micro-pillars are shown in Fig. 3.16b, c, respectively. From observation, the vortex-induced vibration frequency peak in case of the whisker-like design is 50 times smaller than that of the cylindrical design.

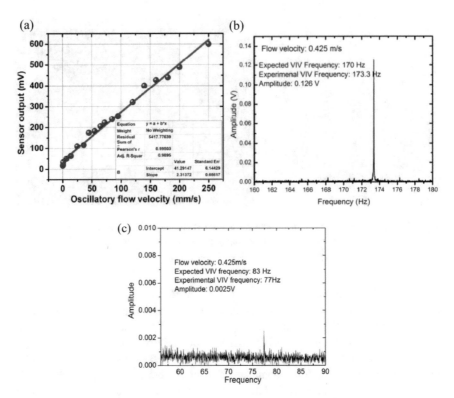

Fig. 3.16 (**a**) Oscillatory flow velocity sensor output. (Figure reproduced from [42] with permission of IEEE), (**b**) vortex-induced vibration frequency of a cylindrical micro-pillar design, (**c**) vortex-induced vibration frequency of a whisker-like undulated design. (Figures reproduced from [43] with permission of IEEE)

3.3   Conclusions and Future Work

Survival hydrodynamics among underwater animals have long inspired scientists and engineers to develop efficient nature-inspired sensors for sensing tasks in underwater vehicles. In particular, the sensing systems in fishes like SNs and CNs are capable of sensing slightest of changes in pressure and velocity. Using their LLs, fishes are capable of generating a three-dimensional map of their surroundings which help them in navigating and complex maneuvering. In this work, we have reviewed some of the most relevant works focused on underwater creature-inspired biomimetic sensors for various flow sensing applications.

The mechanoreceptors of crocodiles can be an inspiration for developing underwater passive sensing for perceiving surrounding flows and disturbances of AUVs. The dome-shaped ISOs in crocodiles have evolved to be highly efficient receptors to sense both the direction and magnitude of flows and oscillating disturbances.

The efforts to replicate these mechanoreceptors as the sensing elements for AUVs to perceive all flows in its vicinity have shown encouraging results. By strategically arranging the artificial SA and RA domes on the surface of an AUV, an optimal sensing strategy to detect the surrounding flows would lead to a more efficient system. The work done serves as the basis for future developments to carry out quantitative analysis to establish a correlation between the stimulus and sensitivity of the dome-shaped sensors. This analysis can be useful in predicting the direction and the source of stimulus based on artificial intelligence algorithms. In doing so, the resolution of directionality detection needs to be enhanced to achieve this objective.

Taking inspirations from seal vibrissae have led several researchers in developing artificial whisker-like sensory systems to mimic the highly sensitive whiskers of the seals to detect water disturbances and track hydrodynamic trails. Various methods and architectures have been proposed, and without a doubt, each has its own advantages and drawbacks. Further work in optimizing the geometry of the artificial whiskers to reduce vortex-induced vibrations needs to be investigated to enhance the sensitivity in low flow rates. Although the underlying principles of the sensory systems in animals are not fully understood, scientists and engineers have begun to explore the mechanisms supporting its advanced capabilities and to distinguish the features of incoming signal characteristics that support wake detection similar to the wake information received and used by the seals. With this information, further development of the hydrodynamic sensory systems can achieve much more than just basic detection of hydrodynamic events.

Acknowledgments This research is supported by the National Research Foundation (NRF) Singapore under its Campus for Research Excellence and Technological Enterprise program. The Center for Environmental Sensing and Modeling (CENSAM) is an interdisciplinary research group of the Singapore-MIT Alliance for Research and Technology (SMART).

References

1. Triantafyllou, M. S., Weymouth, G. D., & Miao, J. (2016). Biomimetic survival hydrodynamics and flow sensing. *Annual Review of Fluid Mechanics, 48*, 1–24. https://doi.org/10.1146/annurev-fluid-122414-034329.
2. Windsor, S. P., Norris, S. E., Cameron, S. M., et al. (2010). The flow fields involved in hydrodynamic imaging by blind Mexican cave fish (Astyanax fasciatus). Part I: Open water and heading towards a wall. *The Journal of Experimental Biology, 213*, 3819–3831. https://doi.org/10.1242/jeb.040741.
3. Windsor, S. P., Norris, S. E., Cameron, S. M., et al. (2010). The flow fields involved in hydrodynamic imaging by blind Mexican cave fish (Astyanax fasciatus). Part II: Gliding parallel to a wall. *The Journal of Experimental Biology, 213*, 3832–3842. https://doi.org/10.1242/jeb.040790.
4. Soares, D. (2002). An ancient sensory organ in crocodilians. *Nature, 417*, 241–242. https://doi.org/10.1038/417241a.
5. Dehnhardt, G., Mauck, B., Hanke, W., & Bleckmann, H. (2001). Hydrodynamic trail-following in harbor seals (Phoca vitulina). *Science (80-), 293*, 102–104. https://doi.org/10.1126/science.1060514.

6. Fish, F. E., Howle, L. E., & Murray, M. M. (2008). Hydrodynamic flow control in marine mammals. *Integrative and Comparative Biology, 48*, 788–800.
7. Yanase, K., Herbert, N. A., & Montgomery, J. C. (2012). Disrupted flow sensing impairs hydrodynamic performance and increases the metabolic cost of swimming in the yellowtail kingfish, Seriola lalandi. *The Journal of Experimental Biology, 215*, 3944–3954. https://doi.org/10.1242/jeb.073437.
8. Yanase, K., & Saarenrinne, P. (2015). Unsteady turbulent boundary layers in swimming rainbow trout. *The Journal of Experimental Biology, 218*, 1373–1385. https://doi.org/10.1242/jeb.108043.
9. Montgomery, J. C., Baker, C. F., & Carton, A. G. (1997). The lateral line can mediate rheotaxis in fish. *Nature, 389*, 960–963. https://doi.org/10.1038/40135.
10. Montgomery, J. C., Coombs, S., & Baker, C. F. (2001). The mechanosensory lateral line system of the hypogean form of Astyanax fasciatus. *Environmental Biology of Fishes, 62*, 87–96.
11. Montgomery, J., Coombs, S., & Halstead, M. (1995). Biology of the mechanosensory lateral line in fishes. *Reviews in Fish Biology and Fisheries, 5*, 399–416. https://doi.org/10.1007/BF01103813.
12. Bora, M., Kottapalli, A. G. P., Miao, J., & Triantafyllou, M. (2017). Sensing the flow beneath the fins. *Bioinspiration & Biomimetics*. https://doi.org/10.1088/1748-3190/aaa1c2.
13. DIJKGRAAF, S. (1963). THE FUNCTIONING and SIGNIFICANCE OF THE LATERAL-LINE ORGANS. *Biological Reviews, 38*, 51–105. https://doi.org/10.1111/j.1469-185X.1963.tb00654.x.
14. Peleshanko, S., Julian, M. D., Ornatska, M., et al. (2007). Hydrogel-encapsulated microfabricated haircells mimicking fish cupula neuromast. *Advanced Materials, 19*, 2903–2909. https://doi.org/10.1002/adma.200701141.
15. Anderson, K. D., Lu, D., McConney, M. E., et al. (2008). Hydrogel microstructures combined with electrospun fibers and photopatterning for shape and modulus control. *Polymer (Guildf), 49*, 5284–5293. https://doi.org/10.1016/j.polymer.2008.09.039.
16. McConney, M. E., Chen, N., Lu, D., et al. (2009). Biologically inspired design of hydrogel-capped hair sensors for enhanced underwater flow detection. *Soft Matter, 5*, 292–295. https://doi.org/10.1039/B808839J.
17. Kottapalli, A. G. P., Bora, M., Asadnia, M., et al. (2016). Nanofibril scaffold assisted MEMS artificial hydrogel neuromasts for enhanced sensitivity flow sensing. *Scientific Reports, 6*, 19336. https://doi.org/10.1038/srep19336.
18. Kottapalli, A. G. P., Asadnia, M., Miao, J., & Triantafyllou, M. (2015). Soft polymer membrane micro-sensor arrays inspired by the mechanosensory lateral line on the blind cavefish. *Journal of Intelligent Material Systems and Structures, 26*, 38–46. https://doi.org/10.1177/1045389X14521702.
19. Asadnia, M., Kottapalli, A. G. P., Miao, J., et al. (2015). Artificial fish skin of self-powered micro-electromechanical systems hair cells for sensing hydrodynamic flow phenomena. *Journal of the Royal Society, Interface, 12*, 20150322. https://doi.org/10.1098/rsif.2015.0322.
20. Coombs, S. (2001). Smart skins: Information processing by lateral line flow sensors. *Autonomous Robots, 11*, 255–261. https://doi.org/10.1023/A:1012491007495.
21. Windsor, S. P., & McHenry, M. J. (2009). The influence of viscous hydrodynamics on the fish lateral-line system. *Integrative and Comparative Biology, 49*, 691–701. https://doi.org/10.1093/icb/icp084.
22. McConney, M. E., Anderson, K. D., Brott, L. L., et al. (2009). Bioinspired material approaches to sensing. *Advanced Functional Materials, 19*, 2527–2544. https://doi.org/10.1002/adfm.200900606.
23. Coombs, S., & Van Netten, S. (2005). The hydrodynamics and structural mechanics of the lateral line system. *Fish Physiology, 23*, 103–139.
24. Tao, J., & Yu, X. B. (2012). Hair flow sensors: From bio-inspiration to bio-mimicking—A review. *Smart Materials and Structures, 21*, –113001. https://doi.org/10.1088/0964-1726/21/11/113001.

25. Leitch, D. B., & Catania, K. C. (2012). Structure, innervation and response properties of integumentary sensory organs in crocodilians. *The Journal of Experimental Biology, 215,* 4217–4230. https://doi.org/10.1242/jeb.076836.

26. Dehnhardt, G., Mauck, B., & Bleckmann, H. (1998). Seal whiskers detect water movements [6]. *Nature, 394,* 235–236.

27. Williamson, C. H. K., & Govardhan, R. (2004). VORTEX-INDUCED VIBRATIONS. *Annual Review of Fluid Mechanics.* https://doi.org/10.1146/annurev.fluid.36.050802.122128.

28. Beem, H., Liu, Y., Barbastathis, G., & Triantafyllou, M. (2014). Vortex-induced vibration measurements of seal whiskers using digital holography. Ocean 2014 – Taipei. https://doi.org/10.1109/OCEANS-TAIPEI.2014.6964469.

29. Hanke, W., Witte, M., Miersch, L., et al. (2010). Harbor seal vibrissa morphology suppresses vortex-induced vibrations. *The Journal of Experimental Biology, 213,* 2665–2672. https://doi.org/10.1242/jeb.043216.

30. Beem, H. R., & Triantafyllou, M. S. (2015). Wake-induced "slaloming" response explains exquisite sensitivity of seal whisker-like sensors. *Journal of Fluid Mechanics.* https://doi.org/10.1017/jfm.2015.513.

31. Yang, Y., Chen, J., Engel, J., et al. (2006). Distant touch hydrodynamic imaging with an artificial lateral line. *Proceedings of the National Academy of Sciences of the United States of America, 103,* 18891–18895. https://doi.org/10.1073/pnas.0609274103.

32. Chen, N., Tucker, C., Engel, J. M., et al. (2007). Design and characterization of artificial haircell sensor for flow sensing with ultrahigh velocity and angular sensitivity. *Journal of Microelectromechanical Systems, 16,* 999–1014. https://doi.org/10.1109/JMEMS.2007.902436.

33. Bora, M., Kottapalli, A. G. P., Miao, J. M., & Triantafyllou, M. S. (2017). Fish-inspired self-powered microelectromechanical flow sensor with biomimetic hydrogel cupula. *APL Materials, 5.* https://doi.org/10.1063/1.5009128.

34. Paull, L., Saeedi, S., Seto, M., & Li, H. (2014). AUV navigation and localization: A review. *IEEE Journal of Oceanic Engineering, 39,* 131–149. https://doi.org/10.1109/JOE.2013.2278891.

35. Scalabrin, C., Marfia, C., & Boucher, J. (2009). How much fish is hidden in the surface and bottom acoustic blind zones? *ICES Journal of Marine Science, 66,* 1355–1363.

36. Kanhere, E., Wang, N., Kottapalli, A. G. P., et al. (2016). Crocodile-inspired dome-shaped pressure receptors for passive hydrodynamic sensing. *Bioinspiration & Biomimetics, 11.* https://doi.org/10.1088/1748-3190/11/5/056007.

37. Stocking, J. B., Eberhardt, W. C., Shakhsheer, Y. A., et al. (2010). A capacitance-based whisker-like artificial sensor for fluid motion sensing. In *Proceedings of IEEE Sensors* (pp. 2224–2229). Kona, HI, USA: IEEE.

38. Eberhardt, W. C., Shakhsheer, Y. A., Calhoun, B. H., et al. (2011). A bio-inspired artificial whisker for fluid motion sensing with increased sensitivity and reliability. In *Proceedings of IEEE Sensors* (pp. 982–985). Limerick, Ireland: IEEE.

39. Valdivia, Y., Alvarado, P., Subramaniam, V., & Triantafyllou, M. (2012). Design of a bio-inspired whisker sensor for underwater applications. In *Proceedings of IEEE sensors.* Taipei, Taiwan: IEEE.

40. Alvarado, P. V., Subramaniam, V., & Triantafyllou, M. (2013). Performance analysis and characterization of bio-inspired whisker sensors for underwater applications. In *IEEE International Conference on Intelligent Robots and Systems* (pp. 5956–5961). Tokyo, Japan: IEEE.

41. Eberhardt, W. C., Wakefield, B. F., Murphy, C. T., et al. (2016). Development of an artificial sensor for hydrodynamic detection inspired by a seal's whisker array. *Bioinspiration & Biomimetics, 11,* 056011. https://doi.org/10.1088/1748-3190/11/5/056011.

42. Kottapalli, A. G. P., Asadnia, M., Hans, H., et al. (2014). Harbor seal inspired MEMS artificial micro-whisker sensor. In *Proceedings of the IEEE international conference on micro electro mechanical systems (MEMS).* San Francisco, USA: IEEE.

43. Kottapalli, A. G. P., Asadnia, M., Miao, J. M., & Triantafyllou, M. S. (2015). Harbor seal whisker inspired flow sensors to reduce vortex-induced vibrations. In *Proceedings of the IEEE international conference on micro electro mechanical systems (MEMS).* Estoril, Portugal: IEEE.

Index

© The Author(s), under exclusive licence to Springer Nature Switzerland AG 2019
A. G. P. Kottapalli et al., *Self-Powered and Soft Polymer MEMS/NEMS Devices*,
SpringerBriefs in Applied Sciences and Technology,
https://doi.org/10.1007/978-3-030-05554-7

Printed in the United States
By Bookmasters